Gold Placer Mining in California

California Bureau of Mines

Compiled by the staff of
the California Bureau of Mines

with an introduction by Kerby Jackson

Introduction

It has been ninety years since the Department of Interior released it's important publication "Placer Mining in California". First released in 1923, this important volume has now been out of print for nearly a century and has been unavailable to the mining community since those days, with the exception of expensive original collector's copies and poorly produced digital editions.

It has often been said that "*gold is where you find it*", but even beginning prospectors understand that their chances for finding something of value in the earth or in the streams of the Golden West are dramatically increased by going back to those places where gold and other minerals were once mined by our forerunners. Despite this, much of the contemporary information on local mining history that is currently available is mostly a result of mere local folklore and persistent rumors of major strikes, the details and facts of which, have long been distorted. Long gone are the old timers and with them, the days of first hand knowledge of the mines of the area and how they operated. Also long gone are most of their notes, their assay reports, their mine maps and personal scrapbooks, along with most of the surveys and reports that were performed for them by private and government geologists. Even published books such as this one are often retired to the local landfill or backyard burn pile by the descendents of those old timers and disappear at an alarming rate. Despite the fact that we live in the so-called "Information Age" where information is supposedly only the push of a button on a keyboard away, true insight into mining properties remains illusive and hard to come by, even to those of us who seek out this sort of information as if our lives depend upon it. Without this type of information readily available to the average independent miner, there is little hope that our metal mining industry will ever recover.

This important volume and others like it, are being presented in their entirety again, in the hope that the average prospector will no longer stumble through the overgrown hills and the tailing strewn creeks without being well informed enough to have a chance to succeed at his ventures.

Kerby Jackson
Josephine County, Oregon
June 2014

GOLD PLACERS OF CALIFORNIA.

Foreword.

In spite of the fact that the gold placers of California have produced over a billion dollars since their discovery in 1848, the idea that they are now completely exhausted is entirely erroneous. The closing down of the hydraulic mines of the State was accomplished by the famous Sawyer decision of 1884, and their attempted reopening, under the terms of the Caminetti Act of 1893, was so bitterly fought at every turn by the agricultural interests of the State, that the industry, so far as tributaries of the Sacramento and San Joaquin rivers are concerned, has lapsed into a moribund condition.

The primary cause of this was the complete disregard of the rights of the farmers of the State by certain mining interests prior to 1884 and a revengeful spirit shown by certain farming interests toward the miners when the agricultural interests gained the upper hand in the courts.

Had a spirit of compromise prevailed in the ranks of both the mining and the agricultural interests, there would have been in 1884, and there would be now, no reason why a workable plan for the continuation of hydraulic mining should not have been put into effect. In the course of an investigation of placer mining conditions in this State, made by the California State Mining Bureau, in the past two years, certain facts have been clearly established; and it is the purpose of this report to prove these facts, which may be summarized as follows:

The principal gold placer area of California lies in the Sierra Nevadas between Susanville on the north and Mariposa on the south. This area is the one that is restricted by the present debris law. It is tributary to the Sacramento and San Joaquin rivers, which have been classed as navigable streams. Aside from this area, there is a very considerable yardage of available gravel still remaining on the tributaries of the Klamath River, which will be discussed in the economic section of this report. The district within the Sierras, however, is by far the most important from an economic standpoint. From investigations thus far, it seems fairly safe to assume a total of about seven billion yards of gravel distributed among the different drainage areas, as follows approximately: Feather River 500,000,000 yards; Yuba River 3,500,000,000 yards; Bear and American rivers 2,500,000,000 yards; Mokelumne, Cosumnes, Calaveras, Tuolumne, and Stanislaus rivers 500,000,000 yards. The above figures include both drift and hydraulic ground.

The above total of 7,000,000,000 yards is of course not all available for working from an economic standpoint. It is fairly safe to assume that at least 40 per cent of this, due to its location and elevation with regard to available water, is not feasible for working. This leaves, roughly, about 4,000,000,000 yards, which, judging from past performance in the days of unrestricted mining, should yield an average of about 15 cents a yard; and if hydraulic mining is restored under proper restrictions with regard to controlling of debris, there is, roughly in the neighborhood of $600,000,000 to be recovered from these drainage systems alone.

A perfectly feasible plan for the working of this ground, under the provision of the Caminetti Act, is now suggested and is to some extent now being carried out by private corporations. Should the work be amplified to cover the whole drainage system, it should properly be under the control of the national and state governments in conjunction.

Briefly, the plan would consist of the erection of concrete debris dams at controlling points on the Feather, the Yuba, the Bear, the American, the Cosumnes, the Mokelumne, and the Calaveras rivers. These dams will take care, not only of the hydraulic miners' debris, but also of the natural erosion, which is now being carried into the Sacramento and San Joaquin rivers by their tributaries, regardless of the restriction and even elimination of hydraulic operations.

By various independent investigators it has been estimated that the debris carried down from natural erosion is from 100 to 200 per cent greater than the amount caused by the mining operations.

These concrete dams will take care of all of the debris, with the exception of such slimes and flocculent matter as are carried over them in suspension by the winter and spring floods. This material can readily be taken care of by the erection of barriers in the Sacramento and San Joaquin valleys similar to the Yuba barrier at Daguerre Point. The material impounded behind these barriers can be pumped out and used for the reclaiming of swamp lands in the vicinity at a cost not to exceed five cents a cubic yard.

The advantages to the State of California which would accrue from a construction of these works would be tremendous. In the first place, the Sacramento and San Joaquin rivers, notably the former, would begin cutting down to the old grades, which existed in 1848. This would result in the restoration of navigation in these rivers up to Colusa and Marysville, as in former days. The advantage to farmers of convenient transportation on the Sacramento River would be of great value.

The effect of this clearing of the rivers would first be felt in San Francisco Bay. The action of the tidal flow in the Golden Gate would in a few years clear the channel to Mare Island Navy Yard and would prevent the abandonment of this navy station, which is now being seriously considered by the government, due to the continual silting up of the channel from the debris now being carried down by natural erosion. Proceeding up the rivers, the expense of dredging and keeping the channel clear on both the Sacramento and the San Joaquin would gradually be eliminated. The necessity of building levees would be gone in a few years, and the direct expenditure saved to the taxpayers of this State and those of the nation would be considerable. The fact of easily available transportation of the farming products of the Sacramento Valley to centers of distribution would result in lowering the cost of farm products to the consumer.

The miner, secure in the knowledge that his tailings were being impounded permanently and without any injury whatsoever to the farming lands below, would begin storing water for mining operations in every available creek and river. This water, which now goes to waste in the annual winter and spring floods, especially during the months of April and May, would be held back in small reservoirs and turned into the river during the months of July and August, when it is most needed by the farmers below for irrigation purposes. This in itself would be worth thousands, and even millions of dollars, to the farmers.

In cutting down to their old grades, the Sacramento and its tributaries would automatically reclaim several hundred thousand acres of land which is now swamp, and which would be converted by this action into rich farming land, thereby changing its value from about ten dollars an acre to from ten to thirty times that much. The counties that would be notably and favorably affected by this action would be Colusa, Sutter, Yuba, Glenn, and Yolo. The increased value of land due to this reclamation would in itself pay for the cost of installation of the projected dams.

The power companies would benefit materially from the government construction of these dams. At present the cost of these works is defrayed by bonding the power companies, and service charges made are ultimately borne by the consumer to carry the interest and principal of these bonds. If the cost of these dams was initially borne by the government and they were leased for power purposes on long term leases to the present operating power companies at a figure which would carry their amortization over a period of, say, one hundred years, the increased power available could be furnished under federal and state regulation to the consumers of California at a very moderate figure and still leave a fair profit for the companies. In addition to this, the power companies would rest secure against any fundamental invasion of their future possibilities of power development from freak legislation of any sort or type whatsoever.

The advantages to the cities of Sacramento and Marysville, as well as many other communities, from the flood protection offered by these dams, is incalculable. From the earliest days of the civilized occupation of California, about once in a generation an uncontrollable flood has resulted in the loss of millions of dollars and of many lives. These floods are periodically recurrent. Under present conditions, whenever the flood plane of the American River and that of the Yuba and its tributaries rise concurrently and not separated by from four to five days, as is generally the case, the cities of Sacramento and Marysville are doomed to a loss which, in their present condition, would run from five to fifty millions of dollars, depending upon the severity and extent of the flood. No levees as yet constructed or capable of being constructed will control the Sacramento River under flood conditions, for the reason that the cross sectional area of the Carquinez Straits will discharge only one-fifth of the volume of water that the tributaries of the Sacramento and San Joaquin rivers will pour down under maximum flood conditions. The excess water backs up in the valley and at stated intervals, generally about once in a generation, will completely inundate the two cities above mentioned and cause uncontrollable loss.

It has been proved in the case of the city of Dayton, Ohio, that the building of control dams on the tributaries of the river is the only means of avoiding such a catastrophe. By proper supervision and emptying of the reservoirs behind the dams upon the approach of such a flood, the maximum discharge into the rivers can be controlled so as to maintain a steady flow, which will not materially endanger the cities along the banks of the river.

The above are the advantages to the people of the State not directly affected by the gravel miner. It is not proposed to restore hydraulic mining on the scale in which it was practiced during the days from 1870 to 1884. There is gravel available in the Sierras to keep the hydraulic miner going for the next hundred years at an annual production of five million dollars and over. This production will result in an increased value given to taxable property in all the mountain counties of the Sierras. Many thousands of men will be given employment, not only in the construction of these dams, but in the working of the gravel properties. The quartz mines themselves will be helped from the reduction in their taxation, from the increased available supply of labor, and from the fact that during the summer and fall months innumerable prospectors who have been working in the placer mines during the winter and spring, will be spread over the country and will undoubtedly discover and begin to open new quartz properties for development by capital later on.

Every merchant in every small town, village and city in California will benefit by the introduction of this flood of new gold every year. Their sales will be increased by the greater buying power of the mountain counties.

Under the provisions of the Caminetti Act, these dams are provided for, their original cost to be borne by the government and to be repaid by a three per cent income tax borne by the miners. As a matter of fact, this cost will be repaid many times over by the value of the reclaimed land, the water available to the farmers for irrigation, the power available to the consumers through the power companies, the reduced cost of transportation, and the restoration of the old, open clear channels to the State of California.

So far as the miner is concerned, his benefit is the smaller consideration of this plan, as his increased scale of operation and production, resulting in an increase of five to ten millions per year in gold of the State, is very small, compared to the value of the increased production and lowering in cost of the agricultural products of the valley.

In conclusion, it may be stated that the purpose of this report is to prove the facts that have been stated in the above summary. To this end the report has been divided into an economic chapter showing the resources and availability of the placer gravels of the State; a chapter showing the most profitable methods of working them; and a chapter showing the feasibility and going into the details of the plan for the restoration of hydraulic mining which has been outlined above.

Again it should be emphasized that this plan will work primarily for the agricultural interests of this State, and secondly for the benefit of the miners, and through them to the benefit of the power companies, the merchants, the artisans and mechanics; and in short, to the benefit of the people of the State as a whole. There is no single individual in the State who works with hand or brain who will not benefit by the increased agricultural and mineral production that will be poured into the coffers of the State for hundreds of years in the future as a result of the measures recommended above; and the increased prosperity of the State will be reflected in that of the country at large.

CHAPTER I.

CONTROL OF MINING DEBRIS.

As the early history of the American occupation of California is associated with the gold miners of 'forty-nine,' so the economic development of the State is bound up with the history of hydraulic mining. During the period from 1848 to 1860, the working of the gravels by hand was very easy, owing to the thousands of acres of light gravel whose shallow banks held the concentrates of the gold derived from the ancient eroded auriferous channels. This work in the early days was done chiefly by the use of the pan, the rocker, the long tom, and other crude gold-saving devices. Later, the ground sluice was developed, and to aid in the washing of gravel through this, water was applied under pressure with cotton and rubber hoses through nozzles against the gravel banks.

Later, iron pipe was used in place of hose, and the nozzle was increased in size and changed in form until the hydraulic giant was evolved. From this point on, the development of hydraulic mining was very rapid, and in 1876, when it had obtained its maximum of growth, there was over one hundred million dollars invested in plants, equipment, and property, and the annual yield was from eleven millions to thirteen millions of dollars. Compared with the fifty millions to eighty millions of the annual yield before 1857, this does not seem very large, but without it, during the decade of 1870 to 1880, the annual yield of California gold would have been insignificant.

In the following résumé free use is made by the author of material collected by him in an article on this subject, published in the issues of 'Mining and Scientific Press' of San Francisco on December 12 and 19, 1914.

While the hydraulic mines were flourishing, the owners apparently overlooked the growth of another industry which was one day destined to drive them from their own field by the action of the courts. The broad valleys, where the tributaries of the Sacramento and San Joaquin unite with these streams, teemed with fertility and these rich farming lands soon attracted the attention of immigrant farmers. Land, which the miners could originally have bought for a song and held as a perpetual dumping ground, steadily mounted in value and its increasing productivity permitted the farming interests gradually to outgrow the mining in economic importance.

The accumulation of debris in the rivers was long a source of annoyance and considerable damage to the farmers. Even during the days of ground sluicing, the great flood in the winter of 1862 covered the richest bottom and orchard lands along the Bear and Yuba rivers with tons of debris from the gold washing. At this time, however, mining was regarded as the more legitimate and powerful industry, and but little protest was made. As the strength of the agricultural interests grew, protests against the overloading of the rivers became more and more frequent and more powerful until the struggle culminated in the famous Sawyer decision on January 23, 1884. This decision, in the form of an injunction, handed down by the United States Circuit Court, in the case of *Edwards Woodruff* vs. *North Bloomfield Gravel Mining Com-*

pany et al., wiped out at one blow property values exceeding one hundred million dollars and indefinitely postponed the addition to the world's wealth of what has been roughly estimated at from five hundred million dollars upwards in the value of placer gold.

The Sawyer decision was, however, but the culmination of a long struggle. The farmers of Sutter, Yuba, Sacramento, and a few other of the valley counties, organized in the early seventies the Anti-Debris Association, to take their struggle into the courts. They were opposed by the once powerful California Miners' Association, and the struggle was long and bitter. The story of the fight from the miners' point of view was especially interesting, and from it one can gain an idea of the principles for which each side was contending. To that end, I quote from a circular issued by the Miners' Association on June 15, 1883, which briefly reviews the then existing situation.

* * * I deem it proper to revert to the services of this association in the defence of the mining industry. The first case of importance was that of *Keyes* vs. *Little York* and 38 other defendants, begun July 29, 1876. After a long trial, wherein the best legal and engineering talent was employed, a judgment was rendered against us. We appealed the case to the Supreme Court of California, where it was reversed on a point raised by demurrer, on the ground of misjoinder of parties defendant.

The City of Marysville then brought suit (September 15, 1879) against the North Bloomfield Gravel Mining Co. and 32 other companies. They included about 200 other persons as defendants. The defendants appealed to the Supreme Court from the injunction issued in that suit and also demurred and applied for change of the place of trial. The motion was refused and an appeal was taken to the Supreme Court, where it is now pending.

* * * Simultaneously with above action (*State of California* vs. *Miocene Mining Co.*) the Attorney General commenced suit (June 28, 1881) against the Gold Run Mining Company of Placer County. Defendants, through the counsel of the association, demurred and answered. The trial came on in October, 1881, and occupied the court 60 days. The presiding judge (Temple) rendered a decision June 12, 1882, sustaining the injunction, with a qualification that on a proper showing of the construction of dams to restrain the coarse material he would entertain a motion to dissolve the injunction. The point raised on demurrer as to want of authority of the Attorney General to use the name of the State, was not decided, but referred to the Appellate Court. Neither party fully acquiesced in the decision. Defendants appealed to the Supreme Court. Appeals were taken by both, and the day of argument is not yet set.

Then the County of Yuba sued the Excelsior Water & Mining Co., the Eureka Lake & Yuba Canal Co. Cons., the Blue Tent Consolidated Hydraulic Gold Mines of California, Ltd., and the Yuba River Gold Washing Co. The last had further time to plead, which is now about expired. Injunctions were issued in all these cases, *ex parte* and without notice to the defendants. Then came the suit of the *County of Sutter* vs. *J. F. Hickey et al.*, commenced April 14, 1882, in the County of Colusa. The defendant's mine is in Nevada County. Also about the same time a suit against the Birdseye-Creek Co. of Placer County.

The last action brought is that of *Edwards Woodruff* vs. *North Bloomfield Gravel Mining Co., Milton Mining & Water Co.* and some eight other companies. It is substantially of the same nature as the others, praying for a perpetual injunction. Woodruff being a citizen of another state the suit was brought in the Circuit Court of the United States, Judge L. B. Sawyer presiding. Defendants demurred on ground of misjoinder of parties defendant. The demurrer was argued by the ablest counsel in the state, occupying the court two weeks, and was overruled. The testimony is now being taken before commissioners for the main trial in September.

All of the above actions have been defended by the Miners' Association, at an aggregate expense of not less than $200,000.

The decision in the case last referred to may be quoted in its principal part as follows:

And that the defendants herein * * * and each and all of their servants, agents and employees, are perpetually enjoined and restrained from discharging or dumping into the Yuba River, or into any of its forks or branches, or into any stream tributary to said river or any of its forks or branches, and especially into Deer creek, Sucker Flat ravine, Humbug creek, or Scotchman's creek, any of the tailing, boulders, cobble stones, gravel, sand, clay, debris, or refuse matter from any of the tracts of mineral land or mines described in the complaint. And also from causing or suffering to flow into said rivers, creeks, or tributary streams aforesaid therefrom, any of the tailing, boulders, cobble stones, gravel, sand, clay, or refuse matter resulting or arising from mining thereon. And also, from allowing others to use the water supply of said several mines or mining claims, or any part thereof, for the purpose of washing into said rivers and streams, any earth, rock, boulders, clay, sand, or solid material contained in any placer or gravel ground or mine. * * * That the defendants, or either of them, may, at any time hereafter apply to this court, upon due notice to the complainant * * * for a modification or suspension of this injunction * * * upon any showing which the court may deem sufficient that the conditions have been so changed that the discharge of such mining debris by said parties or party * * * may be resumed or otherwise conducted, so as not to create or continue the nuisance complained of, or a nuisance of similar character, and so as not to injure or damage said complainant, or upon any other grounds hereafter arising satisfactory to the court. And for the purposes aforesaid, the court hereby reserves the power to modify or suspend said injunction in whole or in part, as the exigencies and equities of the case hereafter arising, may require.

The result of the Sawyer decision was far-reaching in its effects. The decree of the court was nominally against the dumping of the debris into streams and rivers tributary to the watershed of the Great Valley; but in actual fact, hydraulic mining was permanently enjoined in all of its major operations. The larger mines were immediately suspended, as injunction after injunction closed them down; and property worth millions went into disuse and decay.

Many of the smaller mines, however, persisted in continuing operations, and the Anti-Debris Association, composed of farmers of the valley, carried on organized opposition to hydraulic mining. Long and costly litigation resulted and continued for years. As injunctions had to be secured in each and every case, the Anti-Debris Association kept men in the field securing evidence, and among the miners feeling ran high against these emissaries of the agricultural interests.

Gradually, however, the operators gave up the struggle. Thousands of people were thrown out of employment, the gold production of the State dwindled, and property values in the mountain counties were decreased by millions. For about nine years that condition continued in spite of the fact that the Miners' Association used every effort to mitigate the situation from the standpoint of the miners.

A review of the situation, already growing serious in 1881, by the Miners' Association, sets forth the losses resulting from the closing down of hydraulic mining in very emphatic fashion. Following are some extracts from this review:

California has produced (1848-1880) * * * between $1,100,000,000 and $1,200,000,000 of gold, of which very much the larger portion has come, either directly or indirectly, from the ancient or pliocene river channels. * * * As the (surface) placers became exhausted, the miners naturally turned their attention to the sources from which the placer gold had been derived. Commencing in 1851, they have steadily prosecuted the development and working of the ancient river beds, until the yield from them amounts annually (1881) to a sum varying from $11,000,000 to $13,000,000, with the prospect for many years to come of equally great returns. In order to accomplish this result, the expenditure of large sums of money has been required in building dams, canals, and tunnels. The Bloomfield and Milton companies afford a good illustration of the large capital necessary for the successful development of such properties. Work was vigorously commenced on the mines of these companies in 1854, since when about $4,000,000 has been expended—all representing capital account—until 1878, when their works were finally completed. All this $4,000,000, with unimportant exceptions, was furnished by stockholders resident in California.

We roughly estimate the present actual value of these mines in California to be $80,000,000; adding to this the property whose value is dependent on the existence of these mines, there results a total present value of probably $100,000,000.

These statements, made in 1881, show how strongly the danger of extinction of the hydraulic mines was felt at that time. They also serve to emphasize what a tremendously important effect upon the economic progress of California hydraulic mining has had. To continue:

In the early days of placer mining, when there were literally hundreds of thousands of miners washing the soil in every gulch, ravine, and river of the Sierras, the aggregate quantity of material washed into the rivers must have exceeded the amount which by improved, but similar appliances, is now being placed in them. * * * The gravel channels of California have yielded in the past nearly $1,000,000,000 of gold, which has been a mighty force in bringing about the existing prosperity of the civilized world. A still greater quantity remains in our unworked but developed channels, ready for extraction. * * * Can California—can the United States—suffer this great treasure to be forever locked up in our mountains?

Extent of Damage.

The State Engineer in January, 1880, after detailed and careful examination, reported that 43,546 acres of land had been depreciated in value by the flow of mining debris, with a resulting damage amounting to $2,597,635. * * * That the Feather and Sacramento rivers have been injured is conceded by all, but we have the official statement of Captain Eads—as competent an authority on such matters as can be found in the world—that if the flow of heavy sands is kept from entering those streams, they can easily, by proper treatment, be brought into excellent condition. * * * It seems to be taken for granted that we have to deal with engineering problems more difficult of solution than have before been encountered. This is not the case, for both in France and Italy many of the rivers carry larger quantities of earthy material, in proportion to the water, than does the Sacramento. In these countries no great difficulty has been experienced in protecting the lower lands and rivers.

From the examinations and reports of our engineers, we became satisfied, several years since, that it was practicable, at a comparatively moderate expense, safely to store for many years to come all the injurious flow of mining debris in the Yuba, Bear, and American rivers by forming reservoirs by the construction of brush and stone dams in the bottoms and cañons of those streams. * * * the now famous 'Debris Act' was strongly advocated by both miners and farmers, who, in that measure, asked for state aid, so that the necessary funds could be secured to establish a thorough system of protection. In this legislation the miners showed their willingness to pay their full share of the burden, by being taxed in three different ways, and much more heavily than any other class. Under the operations of this act two restraining dams were built in 1880, one across the Yuba, and the other across Bear River, at a cost of nearly $200,000. The residue of the money procured by the taxes levied under the authority of the act, amounting to some $290,000, was spent chiefly in building levees, for the purpose of confining the rivers, so that they might, by a scouring action, deepen their beds.

The results of the court decrees have been stated above. For seven or eight years the industry was gradually paralyzed by the activities of the Anti-Debris Association in serving injunctions. In the fall of 1891 a number of miners in Placer County who had suffered from the closing down of the hydraulic mines met and decided to call a county convention to see whether anything could be done to revive the hydraulic mining industry. This convention was called in Auburn with the object of formulating a plan for a state miners' convention and memoralizing congress as to needed legislation for the industry. This convention was later called in San Francisco, and representatives of both the mining and farming interests were invited. The first evidence of harmony of the whole long and bitter struggle was here made manifest, and a common plan was agreed upon by both interests, which, although they were nominally opposed, had begun to realize that they were interdependent one with the other.

The basis of this agreement was the report of the government commission of engineers. This commission was appointed by a special act of congress, upon suggestion of the legislature of California. It appeared from their report that dams and other restraining works could be erected in many of the canyons for the restraining of debris caused by future mining operations, as well as debris already in the rivers from former operations.

The convention of miners then asked congress to accept and adopt the report of the commission and to take steps at once to put into practical operation the plan suggested so that mining might be again carried on under specified conditions, and the debris restrained from the rivers and farming lands. The result of this was the 'Caminetti Act,' introduced into congress in 1892, and passed in 1893. It is under the provisions of this act that whatever hydraulic mining has been done in the restricted area of the State from 1893 until the present day has been carried on. Certain provisions of this act, however, appear to have been overlooked, and it is under these provisions that the present plan for the control of hydraulic mining debris has been worked out and is now presented.

The provisions of this act are, essentially, as follows: A commission is created, known as the California Debris Commission, consisting of three officers of the Engineer Corps of the United States Army, appointed by the President of the United States, with the advice and the consent of the senate. The jurisdiction of this commission, in so far as hydraulic mining is concerned, extends over the territory drained by the Sacramento and San Joaquin river systems. Hydraulic mining which directly or indirectly injures the navigability of these river systems is prohibited and declared unlawful.

It is the duty of this commission to make examinations and surveys, which will improve the rivers and insure them against damage due to mining debris, *natural erosion*, or other causes, with a view of restoring the navigability of these rivers to the condition existing in 1860, *and permitting hydraulic mining to be carried on, provided the same can be accomplished without injury to the navigability of said rivers or injury to the lands adjacent thereto.*

The commission is charged with the making of examinations and surveys of storage sites in the tributaries of these rivers, or in the plains, basins, and swamp lands adjacent to their courses, for the storage of debris or water, with the object of aiding in the improvement and protection of the rivers by preventing the deposition of debris resulting from mining operations, *natural erosion*, or other causes, or for affording relief in flood time and providing sufficient water to maintain a scouring force in the dry season. In addition to this, they are charged with making a study of the hydraulic mining industry, and such research as science and engineering skill may suggest as practicable in devising a *method whereby such mining may be carried on.*

The commission takes note of the condition of the navigable channels and makes its report to the chief of engineers annually, with plans for the construction of the works outlined in this act, together with estimates of the costs thereof.

The owners and operators of hydraulic mines are required by law to comply strictly with the regulations of this commission under an extreme penalty of five thousand dollars fine and one year's imprisonment. In all cases a license or written permission from the commission must be obtained before hydraulic mining can be legally carried on. These licenses are revocable by the commission and will not be given unless their requirements as to restraining barriers or dams are complied with, and unless these dams are properly maintained, and all information asked for by the commission is at all times forthcoming. Licenses may be revoked arbitrarily by the commission.

Hydraulic mining is declared by the act to have the meaning and application customarily given in the State of California. This term embraces all mining operations where water is used under pressure through a nozzle against any bank of earth, gravel, or similar material, thus eroding the bank.

Plans and specifications for the building of all restraining works are submitted for the approval of the commission, and the work is carried on under its supervision. If the commission is of the opinion that these restraining dams are sufficient to protect the navigable rivers from encroachment by debris, the owner of a mine is permitted to commence operations.

In the case of an impounding dam restraining the debris from several properties, the commission fixes the charge for the privilege of dumping therein. The expense of maintaining such dams or works is provided for by this charge. The amount of debris which can be washed is not permitted to exceed the amount which can be impounded within the restraining works erected. The commission may at any time modify its orders or revoke the privilege to mine. Any intentional violation of its orders automatically works a forfeiture of the privileges conferred.

The commission is empowered *to erect dams for the retention of mining debris out of a fund provided for this purpose by a tax of three per cent on the gross proceeds of hydraulic mining operations. This fund is denominated the debris fund, to be expended by the commission in addition to appropriations made by law in the construction and maintenance of such restraining works and settling reservoirs as may be proper and necessary.*

In work done by the United States and the State of California jointly, the commission is empowered to consult with the state engineers, and the result of their findings, if approved by the chief of engineers of the United States Army, is followed by the said commission. The commission is directed and empowered, when sufficient money is deposited in the debris fund, to build at points above the head of navigation in the said rivers and their branches such restraining or impounding dams as may be required to effect the object of clearing the rivers for navigation.

Licenses to mine, under the Caminetti Act, are not transferable and are valid only for the operations of the special individual, company, and property named in the license. An application for license must be advertised in the newspapers to allow any protests to be filed with the commission. After the sites proposed for restraining works have been approved by the commission, authority is given for the construction of the barrier or dam, together with specifications for the work. Any variation in location or change in character of the work from that specified by the commission may cause rejection of the dam. After satisfactory inspection of the completed work by the commission, a revocable license to mine is issued. Until this license is issued, it is illegal to mine. When mining is begun, monthly reports must be submitted to the commission, and in case of any accident to the barrier, mining must cease at once and the commission be promptly notified.

When a dam becomes full of debris, mining must cease until more impounding capacity is provided, either by raising the dam or the construction of new works. This work must be inspected and approved by the commission. Dams must be kept water-tight, and a pool at least two feet deep must be maintained as a settling basin while mining is in progress.

While the above is not the wording of the Caminetti Act exactly, it is, in effect, an abstract of the conditions necessary for the hydraulic miner, in any of the drainage areas affected, to observe in order to continue operations within the law. The power of the Commission is arbitrary. This is, of course, necessary for the efficiency of the act; but it seems a necessary corollary that the men to whom this arbitrary power is entrusted should be thoroughly familiar with the conditions governing the work which they control.

The area embraced under the provisions of the Caminetti Act includes all of the country drained by the Sacramento and San Joaquin rivers and their tributaries. In the northern portion of the State, tributary to the Klamath river system, hydraulic mining has been carried on unrestricted, owing to the extreme grade of the rivers and comparatively small area of agricultural land along their banks.

As a result of the lack of restrictions in this area, hydraulic mining is already within sight of its natural end, due to the exhaustion of profitable ground.

The passage of the Caminetti Act aroused great expectations in the mountain counties. It was regarded as the successful termination of a long and bitter struggle between the mining and agricultural interests of the State. The general idea among the miners was that their former prosperity and flourishing condition would be in a measure restored. It was not expected that mining would be conducted on the same scale as before the restrictive measures were adopted; but it was hoped that two or three millions a year would be added to the decreasing gold production of the State. In the course of a much longer period of time, it was expected that more than five hundred millions of dollars still locked up in the Sierras would be added to the wealth of the country. Another cause for congratulation was that the conditions governing the passage of the act showed that the miners and farmers of the State would act harmoniously and cooperate for the general good. As the act was passed for the purpose of reviving the mining industry, it was rightly expected that the Debris Commission would not assume an attitude that was either hostile or suspicious of the good faith of the miners.

And indeed this seemed to be the case. The members of the Debris Commission bent every effort to serving the best interests of the State. While friendly toward the mining interests, there was still no partiality shown at any point where their interests seemed contrary to the general good of the commonwealth. With this attitude, the miners were well content, and a certain amount of work was resumed. Restraining barriers were built in conformance with the act, and for a time it seemed that all was well.

But this condition was not permanent. Changes came in the Debris Commission, as was natural in a body composed of army officers, in accordance with the exigencies of the service. New men came in, unfamiliar with the requirements of the mining industry, and were inclined to regard it as somewhat of a trespasser.

In addition, the Anti-Debris Association, like many another body originally organized for a worthy purpose, when its purpose was accomplished, speedily degenerated. The employment of scouts, or 'spies' as the miners were disposed to call them, was perhaps the first branch of the activities of the association to show decadence. Petty graft crept in, and men who were disposed to pay for the privilege of working unrestrained, especially if the property in question were out of the way and inaccessible to the officers of the Debris Commission, found out that it was very easy to avoid complaints. On the other hand, men who operated in good faith under the law, if they did not pay tribute to certain individuals, found themselves subjected to continual annoyance and complaint.

This does not in any way reflect on the membership of the Debris Commission. Composed as it is of officers of the United States Army, they are above any suspicion of connivance. On the other hand, they work at considerable disadvantage. In the first place, they serve without extra pay, and while the service is given wholeheartedly, their duties in connection with river and harbor work are so exacting and onerous that they must of necessity work in part on hearsay evidence in deciding where to conduct their investigations. Complaints filed by the Anti-Debris Association being their principal source of information, it followed that the latter body had a certain amount of influence, and its petty influence was embarrassing. Add to this the fact that as soon as a man had served long enough on the Debris Commission to gain a practical idea of the work of the miners, he was apt to be transferred to another post, and it is readily seen that the efficiency of the Commission's work was subject to continual disturbance.

On the other hand, the California Miners' Association, which for many years was the most powerful backer of the hydraulic miners, became, with the development of the quartz interests, less interested in the matter, and devoted itself to the more 'legitimate' forms of mining. Thus the hydraulic miner was left in the lurch, with the result that instead of the tens of millions which were formerly annually produced, and instead of the millions which should now be produced, the present-day production of hydraulic mines in the Sierra region is now figured in a few thousands of dollars.

The loss to the State, under present conditions, is tremendous. In the first place, it has been computed by competent authorities that the loss due to the closing of the mines from 1884 to 1893 was about fifty million dollars in income. From 1893 to the present day, it, of course, greatly exceeds this figure. The actual physical loss in properties may be estimated from the fact that at the time of the Sawyer injunction, there were in the State about 5225 miles of ditch, built at an estimated cost of $3,800 per mile, including cost of reservoirs. This represents an investment of about twenty million dollars. Investment in equipment in the various hydraulic mines of the State total another twenty million dollars. The value of the gravels themselves, judging from the figures at which they were assessed, was about sixty million dollars. This value has been practically wiped out. As a result, it may be said that at one blow the Sawyer decision reduced the visible assets of the State of California by one hundred million dollars.

The loss to the farmers and merchants of the State may be fairly calculated as follows: At the time of the closing down of the hydraulic mines, their annual yield was about five million three hundred thousand dollars. The cost of supplies and labor in the production of this amount was between three and four millions of dollars. Fifty per cent of this was in the form of supplies directly furnished by the farmers of the valley adjacent to the mining region. The merchants of Sacramento, Marysville, Oroville, and other distributing centers, derived considerable benefit from the handling of the supplies consumed by the miners, but the cutting off of this market affected many of them to the point of bankruptcy. In the mountain towns, whole communities of several thousand inhabitants were reduced to villages, and in many cases entire communities were wiped out. This resulted in the depopulation of the mining counties, and a depreciation in assessable values from which they have never recovered.

In addition to the above-mentioned losses, there was a potential loss, which far exceeds all of the others put together, and which directly affected the prosperity of the State of California during the last forty years, and still continues to affect it adversely. According to the figures of the Reclamation Service of the United States Geological Survey, there are about 800,000 acres of unclaimed swamp lands in the Great Valley of California. If the silt and finer gravel sent down by the miners had been diverted to the reclamation of this land, and if the boulders and lighter material had been restrained from the rivers by the construction of restraining dams, under the provisions of the Caminetti Act, the rivers themselves in cutting down to their old grades would have reclaimed a large proportion of these now worthless swamp lands. The actual value of the lands, which are possible of reclamation by an intelligent handling of our river and debris problems, exceeds, at a modest estimate, one hundred millions of dollars.

The reclamation of these lands will result in the addition of a permanent asset to the resources of the State of California, which will be of tremendous value. As a heritage to future generations, these lands, when reclaimed, will support a population almost equal to the present population of the State. Their productive capacity is equal to that of the richest land now existing in the State if they are properly drained and re-soiled with the light debris which passes over the restraining dams built in the higher reaches of the rivers. So far from being a menace to the farmer, a quotation from a resolution, passed by the directors of the irrigation district of Oakdale in May, 1919, will indicate how beneficial this silt is to the lands whose fertility has been exhausted.

The resolutions state that instead of the silt from the rivers injuring the district, it has been highly beneficial, the debris from the river having already greatly enriched the soil. They state that investigations have disclosed that irrigation from the silt-laden river, below the Goodwin dam, has actually built up fertile soil by depositing the silt that came down the river on the land and giving it body, as well as fertility.

In another district, near Turlock, the former chief engineer, Mr. Burton Smith, stated that land that would not even grow sunflowers when the district was organized, now produces enormous crops. Experts ascribe this to the silt deposit on the soil from the rivers.

The above instances are cited from many which have come to the writer's notice since the organization of irrigation districts in the Great Valley of California.

In restraining the operations of hydraulic mining, the need of a training in the principles of the industry would seem to be as necessary as in the actual operation of the mines themselves. The building of satisfactory dams, the recognition of the principles controlling the transportation of sand and clay, as well as light gravel in the material being worked, are all factors which should be taken into account. The principal factor of all, however, and one which under the present law, is a burden resting altogether on the hydraulic miner, is the natural erosion caused by the intermittent flooding of the streams from the torrential storms of the Sierras. The denudation of our forests and the increased cultivation of the Sierra foothills have increased the volume of material borne into the rivers from this source by several hundred per cent.

The work of regulation requires all the time that can be given to it by a body of men specially assigned for this purpose, if arbitrary restrictions are to be removed. There is absolutely no reason why several times the present number of hydraulic mines in the Sierran counties should not be in active operation, so far as the damage done by debris is concerned.

A factor which has tended to curtail the operation of hydraulic mines has been the increasing use of water for power and irrigation purposes, as well as its use for domestic supply for various towns. There is no question but that the power interests have grown to a position of great importance, as well as have the irrigation systems; nevertheless, there is in this fact no need for conflict. During from four to six months of the year, the streams of the Sierras carry an excess volume of water, which all the storage systems yet devised or likely to be devised for some time, can not contain. This water, or a small portion of it, is what the miners would conserve, and instead of running off during the months of April and May, it would be held back and turned into the rivers during the months of July and August, when it is most needed.

The conditions which confront the hydraulic miner today, in spite of the relief supposed to be afforded by the Caminetti Act, are such that the industry is nearly defunct. The remedy for this state of affairs is simple and should be applied; it means the increasing of California's gold production in the neighborhood of five million dollars a year for the next one hundred years or over, and the restoration of a portion of their former wealth to the mountain counties of the State.

In a publication of the United States Geological Survey, entitled 'The Transportation of Debris by Running Water,' there are some very interesting statements. In speaking of the flow of the Yuba River and its suspended load in 1879, when that stream was at its flood and when hydraulic mining was being carried on, the following results were noted: When the discharge was 26,000 cubic feet per second, a load of 0.35 per cent by weight was noted; when the discharge was 18,000 cubic feet per second, the load was 0.42 per cent. At low water in the same year, when the discharge was 510 cubic feet per second, a load of 0.86 per cent was noted. The fact that the river's load is greater in proportion during low water is explained by the fact that the turbid tributaries of the river are less diluted with the clearer water from other tributaries. From these results it will be noted that a decrease of 97 per cent in volume brought about an increase of virtually 150 per cent in the proportionate load.

On the other hand, in 1906, when most of the hydraulic mining was suspended, with a discharge of 33,000 feet, the load was 0.65 per cent. If the increase in load is at all times regular for large volumes (which appears a warrantable assumption), the suspended load for a volume of 18,000 feet, the same as in 1879, would have been about 0.13 per cent, or about one-third of what it was at the time when hydraulic mines were working at capacity. In other words, the volume of debris carried down by natural erosion was, when the mines were mostly shut down, about one-third as much as when they were working full blast. As conditions of stream bed and other factors were in 1879 more favorable to debris suspension than in 1906, and as the amount of intensive farming in the Sierra foothills has vastly increased since 1906, and the area of deforestation has proportionately increased, there is reason to assume that the volume of material annually borne into our rivers by natural erosion is nearly as large as the amount formerly carried into the rivers by hydraulic mining under the old conditions.

The fact that at low water the percentage of debris transported is so much less than it might be calculated to be is, of course, explained by the deposition of debris by the feeble currents and consequent shoaling of the river.

Returning again to official statements, we find the following in the report of the Debris Commissioner, from May, 1901, to December, 1902:

The debris flow has caused the destruction of thousands of acres of fertile farms and the serious impairment of navigable rivers. Its attempted solution by injunctions by the courts has resulted in the destruction of a profitable industry of the state—that of hydraulic mining. Although many reports have been made on the subject, there are certain features that have not been given due weight because the various investigators were not in touch with the actual conditions leading to the debris flow. It is this lack of knowledge that has in a great measure retarded its solution in the past. Careful consideration by those most interested will show that there is common ground to stand on, and it is hoped that it will lead to a 'community of interests' that is so necessary to hasten the much-needed work. * * *

The California Debris Commission in 1894, from the result of borings made by Col. W. H. Heuer, placed the amount of debris from Marysville to Daguerre Point, a distance of 9½ miles, at 308,000,000 cubic yards, and estimated 100,000,000 cubic yards more to the Smartsville dumps, or a total of 408,000,000 cubic yards. I am of the opinion * * * 600,000,000 cubic yards are deposited within this area. Of this amount there were deposited * * * during the period 1879-1899, * * * a total of 100,550,000 cubic yards. This represents the cleaning of the cañons, the washing of the hydraulic and other mines, and from natural causes.

During the past eight years, I have endeavored to determine what the relative erosion was on Deer creek. * * * In the years gone by there was in this watershed, considerable hydraulic mining, which is now practically abandoned. The dumps have been so cut and carried away as not to affect the winter flow. Altogether it can fairly be considered average conditions for natural erosion on the watershed. On January 3 and 4, 1895, there fell 7.08 inches of rain in 23½ hours. The creek was very high and muddy. The percentage of the silt was determined and a rough gaging of the flow made. After making liberal deductions for safety it was found that the creek carried 86,000 cubic yards of material past the town in twenty-four hours. Two-thirds of this, by weight, may be counted as sand, being the mean of a number of observations and checked by comparison of the composition of soils. At that time the Excelsior Water and Mining Company was hydraulicking at Smartsville, impounding the debris in a worked-out pit. It is one of the largest concerns operating under the California Debris Commission. Yet that storm sent down more material in one day than the Excelsior Company mined in six weeks—boulders, cobble, gravel, sand, and 'slickens.'

The point to be emphasized is that with a system of building restraining dams and allowing operation until they are filled up, the gravel miner takes upon himself the burden of restraining the natural erosion of the elements.

By a study of the maps of the United States Geological Survey, it will be noted that the principal Cretaceous and Tertiary gravels of the State will be found to be in the counties of Sierra, Plumas, Nevada, Placer, and El Dorado. Amador, Calaveras, Tuolumne, and Mariposa contain but small quantities of gravel, compared to the immense bodies of the Slate Creek district or the San Juan Ridge, or the Quaker Hill, Red Dog, Gold Run and Dutch Flat region.

During the early days of the industry, work was mainly confined to the lighter surface gravels, and it is from these that most of the mining debris has come. The gravel that it is desired to work at the present day, is as a whole much heavier in character. In gravel of this type, where dumping is not done directly in the rivers, but upon bars adjacent to them, it may be safely estimated that from 75 to 80 per cent of the entire material stays on the original dump or within a few hundred feet therefrom. The finer material and the sand goes down the river to the restraining dams with the exception of what settles between the boulders on the dumps. For this reason, it may be made as a fairly general statement, covering three-quarters of the Cretaceous and Tertiary channels in the Sierras at the present day, that not more than from 20 to 25 per cent of the washed material will come within the field of the restraining dams.

Another point can here be made, which illustrates again the principle of community of interest. Of late years the destruction of valuable farming land by gold dredging operations, has been the subject of a good deal of unfavorable comment. The dredging industry of California is dying a natural death, and there remains little available ground of economic value for dredging purposes. However, as a result of operations already completed, there are in California large areas of once fertile farming and orchard ground which have been reduced to a series of unsightly boulder piles. In Australia, the necessity of reclaiming the damaged lands was long recognized, with the result that in farming communities the gold dredgers were forced at considerable expense to reclaim the land behind them by pumping sand and silt back on top of the boulder piles. In the report of the Secretary of Mines, Victoria, Series 1907-1913, detailed descriptions of the methods adopted, accompanied by photographs, are given. The expense involved in reclaiming was borne by the dredging companies, and notwithstanding the value of the reclaimed land, profits were enormously decreased. Naturally the American companies were loath to follow their example, with the result that certain sections of the State, notably the district around Oroville, remained an eyesore to the beholder. As everybody who has been in the dredging fields knows, these uneven and irregular gravel heaps can be leveled, though at considerable expense, but getting a top soil upon them is another matter. It requires the use of suction pumps to bring up the sand and silt from the river beds. The construction of a few barriers in the lower rivers to restrain such fine material as may be carried over the debris dams, would bring about the reclamation of not only the dredged areas, but a very considerable acreage of barren ground adjacent. A few suction dredges stationed in such places as on the Yuba River near Hammonton, and on the Feather River below Oroville, and behind barriers such as the one at Daguerre Point, could handle all of the excess material that comes over the debris dams and is deposited where the rivers lessen their grades.

In discussing the effects of natural erosion on the rivers of California, it may be noted that in 1897, it was estimated that the debris in the Sacramento Drainage System totaled 1,529,000,000 yards. The total amount of debris originally deposited in the rivers, including the slickens carried off and deposited in the Bay Region, was in the neighborhood of 3,000,000,000 yards. The proportion of this debris, which may be laid at the door of the miners for operations, between 1860 and 1881, is calculated liberally at 650,000,000 cubic yards. This is only 21 per cent of the amount of material deposited in the rivers between 1849 and 1897. The balance of this debris may be charged to the increased area brought under cultivation by the farmers in the foothills; and the operations of the lumber companies in the mountains, which resulted in the deforestation of large areas in the sierras. Among other causes, the erosion brought about by cattle and sheep and by ordinary road and trail traffic, are contributory. It is estimated that the seven thousand miles of road in this area, worn down two feet in depth, would deposit twenty million cubic yards of material.

These figures are further borne out by computation made in Italy on the Po River. There is no hydraulic mining in the drainage of this river, and there never has been. Its drainage area is approximately the same as that of the Sacramento. Through natural erosion alone, it discharges 56,000,000 cubic yards of material annually into the ocean.

To carry out any readjustments of the problem of debris control, certain radical changes would have to be made. It is suggested that in order to do full justice to all of the interests concerned, the work should be under the control, not only of the Debris Commission, composed of officers of the United States Army, but also of a debris commission of experienced engineers, permanently appointed by the State of California. This is in accordance with the provisions of the Caminetti Act. Further, a debris commission, appointed by the Reclamation Service of the United States Geological Survey, in the administration of the reclamation features involved in the problem of debris control, should be, if not in actual control of the work, in close consultation at all times with the other two commissions.

The expense of the building of these restraining and reclamation works should be borne, not only by the miners, but also by the power companies, who would benefit from the opportunity of leasing these dams for power purposes, and by the farmers who occupy the reclaimed lands, a proceeding which is entirely regular and in accord with that of all reclamation projects. Under the provisions of the Caminetti Act, the miners would bear their share by a tax upon their gross income. It is estimated that in carrying out this plan, an expenditure of thirty million dollars would be involved. This fund should be appropriated by the state and national governments in conjunction, and administered under the supervision of the three commissions mentioned above.

The chief advantages of such a readjustment would be: First, a definite understanding between the government and the hydraulic miner as to just where the latter stands; second, a distribution of the burden of restraint which would be just and equitable; third, the removal of the prohibitive tax placed by natural erosion upon the industry of hydraulic mining under present regulations; fourth, a solution of the waste dredge lands problem, and the problem of the reclamation of our eight hundred thousand acres of unproductive swamp lands. Having assumed control of the situation, the joint commission, somewhat along the lines suggested, could take charge of the entire matter and work out the details of a plan, which would be at once satisfactory to the miners and highly beneficial to the agricultural interests of the State. There is plenty of common ground to stand on, and the attitude of the miners and the farmers should be equally liberal. The resumption of hydraulic mining on its former scale is not to be expected, but the six hundred million dollars, more or less, that is at present locked up in the useless ground of the Sierra Nevada, may well be expected in the course of a hundred years to be added to the riches of the commonwealth of California in increasing volume as time goes on and the efficiency of a practical working plan of control is demonstrated.

In the course of the past two years, an investigation made by the writer has resulted in the selection of the sites of six main dams, which would control the principal watersheds of the Sierras.

On the north fork of the Yuba River, below Bullard's Bar, a dam is now being constructed by private interests, which will eventually be 350 feet in height, and will cost upwards of six million dollars. This dam will take care of all the hydraulic mining debris ever likely to be sent into the rivers from any of the hydraulic mines above this point on the north fork.

On the middle fork of the Yuba River, below the junction of Deer Creek, and near Camptonville, a much larger dam with a capacity of half a billion cubic yards is suggested as a piece of work which should be undertaken by the United States Government in conjunction with the State of California. This dam would be about 300 feet in height, and would cost upwards of twelve million dollars.

On the south fork of the Yuba River, below the junction of Shady Creek, there is a good opportunity for a storage dam about 200 feet in height, which would involve an expenditure of two to three millions of dollars. The above dam would take care of the major portion of the debris brought down by the south fork of the Yuba River.

The above are the projected dams for the control of the Yuba River drainage. The estimates, as to cost and storage capacity, are necessarily very rough, as time is not available for a careful examination. The total storage capacity of these three dams sums up to over six hundred million yards. The Bullard's Bar dam on the north fork is already under construction by private interests at an estimated cost of from six to eight million dollars. The total cost to the United States and State governments of the construction of the dams on the middle and south forks, as suggested, would be about fifteen million dollars.

On the Bear and American rivers, tentative selections of dam sites have been made, subject to change upon more extended investigation. On the north fork of the American River, a dam a short ways above the crossing of the Forest Hill-Colfax Road, about 300 feet high, would furnish storage for, roughly, two hundred million yards of tailings at a cost of about seven million dollars. On the middle fork of the American, at the junction of Canyon Creek, a two hundred-foot dam would give storage for over two hundred million yards of tailings. This is a remarkably good storage site, due to the storage space in Otter Creek and the flat grade of the middle fork for four or five miles above this site. The cost of this dam, at the height mentioned, would be roughly around three million dollars.

On Bear River, near Howell Hill, a dam 150 feet in height at a cost of about two million dollars, would give a total storage of about fifty million cubic yards of tailings. The beds of Greenhorn Creek and Bear River are full of tailings from the old hydraulic mines, which have been estimated to run from 20 to 60 cents per cubic yard. The amount of tailings has been estimated at fifty million cubic yards. A diversion tunnel for many years has been planned in the Secret Canyon, which would divert the excess freshet flow into the north fork of the American. This tunnel would be approximately two miles in length. Whether it would be a feasible or practicable plan to work these Bear River tailings off into the American River through a tunnel of this sort for the sake of saving the values in them, is a question that would have to be determined by careful investigation.

In the length of time allowed for this investigation, it has been impossible to make careful estimates of the costs and capacities of the dams mentioned above. They are, however, within reasonable limits, a fairly close approximation. Careful survey made by government engineers would possibly result in more advantageous locations. These dam sites, as suggested, however, control practically the entire dumping space of some three billion yards of workable gravel with a storage capacity of over one billion yards. As the amount of material that would get down to these dams is about one-fifth of the amount that would be worked off, they would probably take care of all material likely to get down to them, in addition to the natural erosion of the next fifty or one hundred years. In time, of course, like all storage dams, they would become filled by natural erosion, and it would be necessary to either raise them or to build others higher up.

In order to take care of the slimes and fine material that would be carried over these dams by the flood waters of winter and spring, it would be necessary to place barriers down in the valleys, where the rivers lose their grade and deposit such fine material as they can not carry to the bay and to the ocean. As a barrier has been constructed at Daguerre Point on the Yuba River, a pump dredge could be installed behind this to remove the fine material as fast as it came down and deposit it over the swamp and unreclaimed land adjacent thereto. On the Feather River, below Oroville, the same plan could be used and much land reclaimed, including a large portion of the waste dredge land. The streams could be widened by diking back, and the greatest amount of suspended matter would thereby be deposited. The slickens could be pumped out, run through sluices for the recovering of such gold as might be obtainable, and used to cover the waste lands with the finest kind of fertile soil.

On the American River, below Folsom, a similar barrier could be erected, and waste lands adjacent thereto could be reclaimed in the same way.

The above plan, while partly tentative in its details, in its general features is absolutely sound. It will inure greatly to the advantage of the farmers of the State by virtue of the reclamation of several hundred thousand acres of waste land; by the storage of water during the spring months, and delivering into the rivers during the dry season, when needed; by permitting the rivers to cut down to their old grade, and giving transportation and navigation to Marysville and Colusa, as in former days. It will also give the farmers a greater market for their product, close at hand, due to the increased prosperity of the mining counties. These advantages to the farmer, far exceed those given to the miners by this plan. Nevertheless, the rejuvenation of hydraulic mining in the mountain counties will result in the restoration of vigorous life to the entire region affected. Towns that have been almost evacuated will again be busy, the buying power of the mountain counties will be greatly increased, and the whole State will reflect their prosperity. Thousands of men will be given employment, and the quartz mining interests will reap advantages from the reduction in taxes and the probability of the opening of new mines by increased prospecting during the summer months.

The State, as a whole, will benefit: first, from the clearing of the Mare Island Navy Yard channel and the bay; second, from the clearing of the lower reaches of the Sacramento, and the avoidance of the present necessity of dredging and diking.

The greatest advantage of all, however, will be gained by the cities of Sacramento and Marysville. About once in every generation, when the American, the Yuba, and the Feather rivers deposit their burden of flood waters into the Sacramento at the same time, a flood occurs, which is absolutely beyond control of any engineering works that may be or are likely to be constructed. This flood occurs for the reason that the Carquinez Straits have only about one-fifth of the cross-sectional area necessary to discharge the entire amount which the Sierra rivers, at flood, will pour into the San Joaquin and Sacramento. By the use of these storage dams and others which may later be built on the Calaveras, Mokelumne, Stanislaus, and Tuolumne rivers, the maximum flood can be held in check by alternately emptying and filling these storage reservoirs in such a way as to distribute the intensity of the flood and to hold back the excess volume of water whenever inundation is threatened. Often the checking of one of these rivers for a few hours is all that is necessary to prevent damage running into thousands, and even millions of dollars. With the increased population and building area of the cities of Sacramento and Marysville, a flood, such as we had in the winter of 1889 and 1890, would be a tremendous calamity.

The advantages to the power companies from the construction of these dams and their leasing by the state and national governments at a figure sufficient to cover a long-time amortization of the costs thereof, will be incalculable. The necessity of heavy bonding on their part would be avoided, and service charges to consumers would be greatly lowered. In this way, the people as a whole would reap the benefit of the enterprise; furthermore, the power companies would be forever secured against the possibility of socialistic and freak legislation, such as has been attempted before, in depriving them of the benefits of their vast expenditures of time and money in the development of the water resources of the State. As the policy of the power companies has always been to distribute their stock and their bonds among the consumers of power and the small investors of the State, this will ultimately result in a permanent benefit to the public at large.

Bibliography.

Mining Debris. Mining & Scientific Press, Jan. 15, 1876; Feb. 12, Mar. 18, Mar. 25, 1876; Mar. 15, 1879; Mar. 22, Nov. 29, 1879; Mar. 20, 1880; July 30, 1881; Aug. 6, Aug. 13, Nov. 12, Dec. 31, 1881; Jan. 22, 1882; July 1, July 29, 1882; June 17, 1882; Jan. 12, 1884; Jan. 19, 1884; May 6, 1893; Feb. 16, 1895; Nov. 14, 1896; Jan. 9, Jan. 16, 1897; Oct. 22, 1898.

Redemption of Great Valley of California. Trans. Am. Soc. Civil Engrs., Vol. 66.

Restraining Barriers in the Yuba River. Mining & Scientific Press, Aug. 16, 1902.

Debris Control. Mining & Scientific Press, Sept. 2, 1905; Dec. 2, 1905; Feb. 16, 1906; Nov. 20, 1909; Nov. 12, 1910.

A New Debris Dam. Mining & Scientific Press, July 10, 1915.

Hydraulic Mining Debris in California. U. S. Geol. Surv., P. P. 105.

Transportation of Debris by Running Water. U. S. Geol. Surv., P. P. 86.

CHAPTER II.

PLACER MINING METHODS.

In investigating a placer deposit with a view to subsequent mining operation, one of the most important things to be considered—aside from the values in the property and the facilities of operation—is the adaptation of methods of mining. Different deposits require different methods of attack, and many mines which might have been successful have failed, due to the use of methods that were not adapted to the property.

PHOTO No. 1.   Characteristic Hydraulicking Gravels in Peru.

PHOTO No. 2.   Close-up of Same Gravels in Peru.

Each individual property requires its own special study of conditions and its own adaptation of methods. This should be done by a trained and experienced engineer; but there are certain general conditions that call for certain methods of mining, and it is proposed in a general way and in language that the ordinary individual interested in placer mining or the prospective investor, may understand, to set forth a view of the best known methods in current practice that have come within the writer's experience.

This chapter does not cover the whole field—the limitations of space forbid—but it is hoped that a careful study of the bibliography attached will serve to cover these points, which are not touched upon in the course of the chapter. In the following résumé, material has been freely used from an article, entitled "The Mining of Alluvial Deposits," written by Newton B. Knox and Charles S. Haley for the "Mining Magazine" of London in 1915.

The most important methods of alluvial mining are dredging, hydraulicking, drifting, mechanical handling, ground sluicing, dry washing, and beach sand work. The last is possibly an adaptation of several of the former methods. Taking up these methods in the order of their importance in modern placer mining, we will begin with dredging.

DREDGING.

The type of deposit that is most suitable for dredging is generally one in a river that has worn down its gradient to a basal plane, as in the case of the Yuba River, above Marysville, California. Another type is that of a mountain stream at the point where it debouches into a main valley and there deposits its burden, as at Oroville, on the Feather River, and at Folsom on the American. A third type is that of a meandering river bordered by the broad flats of its peneplain, as on the Pato River, in Colombia.

An ideal dredging ground should possess the following characteristics: Gravel which is loose and uniformly small, yet sufficiently impervious always to retain sufficient water to float the dredge; a depth which varies from 40 to 100 feet; a bedrock that is soft and decomposed; an absence of barren top soil, beds of sticky clay, cemented gravel, or barren sand; no living or buried timber; and a fairly uniform distribution of value throughout the gravel. The river should not be subject to sudden or extensive floods. In addition to these factors, availability of power, either water or steam, and accessibility to transport, with its consequent low freight rates, should exist. Preferably, the deposit should be situated in the temperate zone.

The gravel should be small, which means for an ideal dredging proposition, the largest stones are smaller than a man's head. Much dredging, however, is done in ground in which the boulders vary in size from a foot in diameter, weighing perhaps 200 to 300 pounds, to boulders weighing two tons and over.

Most of the records of yardage have been made in California on small gravel, but on the Trinity River, operation is now going on in ground that contains boulders weighing up to five tons. As a general rule, unless such a deposit is unusually rich, it is very unwise to attempt to use a bucket dredge; under these conditions repairs to the bucket line are frequent and costly, and the recovery of gold under the larger boulders which lie on the bed rock is very small.

An excess of cemented gravel or stiff clay as a rule results in reduced yardage and expensive repairs. Sticky clay means reduced capacity, due to difficulty in discharging the buckets. It is also a source of considerable annoyance in the sluices, from which it is apt to remove the amalgam. Some troubles due to clay can be overcome by the use of log-washers on the dredge; but extensive beds of sticky clay, even when of a high gold content, as those on the coast of Brazil, are undredgable.

The gravel should be sufficiently water-tight to hold the level of the dredge-pond, because a sudden emptying of the pond would probably overturn the dredge and would wreck not only the hull but the entire mechanical plant. Also it is sometimes necessary to maintain the water in the pond at a different level from that of the surrounding country. Ordinarily from 50 to 75 inches of fresh water is required continually to keep the level of the pond constant and to supply wash water for the digging operations of dredges with capacities running from a 7-foot bucket to a 16-foot bucket.

Deep gravel from 70 to 100 feet is ordinarily worked by dredges ranging from 15 feet in size to 18 feet. This does not mean, however, that the size of the dredge is dependent altogether upon the depth of the gravel. Gravel of a depth of 10 to 20 feet would call for either a very small dredge or for the adoption of another method of working, such as the steam shovel or the drag-scraper. The digging depth of gravel, which means the depth from the surface of the pond to bedrock, is ordinarily considerably less than the total depth of the gravel. This is for the reason that practically all dredges carry a bank considerably above the level of the pond. For instance, in a deposit of extreme depth a large dredge might carry a 30-foot bank and dig to a depth of seventy feet below the water level. This would, of course, call for ideal conditions in the bank.

The ideal bedrock in dredging is one which is soft and without ridges or crevices. In a granite country the bedrock under the gravel is usually decomposed to the required softness. The same thing is true of some sandstones and porphyries. A limestone or slate country is nearly always characterized by a hard, blocky, or knobby bedrock, often with deep crevices. These latter types are unsuited for dredging, as the wear on the bucket lips is excessive, and the gold lodges in crevices beyond their reach. A false bedrock, such as the one of volcanic ash at Oroville, is ideal. The presence of boulders of the country rock of the district is usually an indication of hard bedrock.

In judging the hardness of bedrock, the evidence derived from shafts or drill holes should be accepted in preference to that presented by exposure in running streams or watercourses. In many cases the bedrock exposed in the river will be found to be hard, whereas the drill may prove that the bedrock underlying the gravel adjacent to the stream is soft and decomposed to a considerable depth. This is due to the fact that the denudation caused by the river is greater than the weathering, and all weathering products are swept away as soon as they are formed. Under the gravel, however, all the decomposed minerals may be intact and form a soft bedrock suitable for dredging.

Dredging in tropical countries is often greatly hampered by the presence of buried timbers, as in French Guiana, and sometimes in Victoria. This is always a source of considerable annoyance, and in fact, if they are in great quantity and well preserved, their presence will prohibit successful dredging. Dredges which are built for tropical work should always be equipped with derricks and stump-pullers. In California, gravel is generally free of buried timber, but standing timber is quite common, and unless the gravel is sufficiently deep to distribute the cost of clearing and stumping it over a large yardage, this feature also adds materially to operating costs. A shallow deposit, with a forest of large and deep-rooted trees upon it, will be prohibitive in dredging cost, unless it contains an exceptionally high content of values.

On the other hand, the dense jungle growth of tropical rivers is often misleading, because, while the growth may be impenetrably thick, these

PHOTO No. 3.   Natives Washing with Rocker in Peru.

jungles can be quickly and cheaply cleared, the roots of the larger trees being very close to the surface. In Colombia, where native wages vary from fifty cents to one dollar, an acre of thickly covered ground can be cleared and stumped for $25 to $40.

The gold should be what is known as three-dimension gold, and be readily amalgamated. It should not be too finely divided, and it should be distributed with some uniformity throughout the gravel. If the gold is coarse and confined to bedrock, it is usually a condition of spotty ground, and unless the bedrock is soft, recovery is very difficult. It is very unusual for top soil to carry gold in profitable quantity. Extensive layers of sand are nearly always barren, and a deposit in which much sand occurs is apt to be very erratic in the distribution of its gold content.

Deposits situated in extreme latitudes often contain permanently frozen gravel, which can be dredged only by previous thawing. This thawing is usually done by means of steam-points driven down into the gravel and supplied from central boiler plants through insulated steam lines; or else by what is known as cold water thawing, which is often done by the diversion of large bodies of water over the area to be dredged and also by its forcing into the ground in points under hydraulic pressure. In Alaska frozen ground has been dredged at a cost ranging from 50 to 75 cents per cubic yard. Aside from the question of accessibility, dredging in extreme latitudes is greatly hampered by the short seasons, necessitating either the engaging of a new staff each year or their maintenance during the idle season. Cold climates also necessitate the

PHOTO No. 4. Nechi River at Pato. Characteristic Dredging Gravels in Colombia.

installation of heating apparatus and the housing of the exposed part of the stacker and bucket line.

Dredging in the tropics is likewise more expensive, owing to climatic conditions, with their accompaniment of malaria and other fevers, as most of the dredging is done on the lower river bottoms, where fevers are prevalent. The usual inefficiency of native labor also adds greatly to the cost.

The attempt to dredge in torrential rivers, characterized by sudden and extensive floods, is always a hazardous undertaking. Many dredges have been swept away and wrecked by a sudden rise of the water. The question of floods is one that should be most carefully investigated, and data regarding it should be collected over a period covering a number of years. As a general rule, it may be stated that rivers flowing through narrow canyons are always perilous. Even rivers with adjacent flats. when in countries where cloudbursts are frequent, are liable to floods. Under these circumstances the ground is often worked by dredging inland and keeping a barrier of unworked ground to serve as protection.

The importance of freight rates weighs more heavily with this class of alluvial installation than any other. For instance, a 7-foot, close-connected dredge of the California type, complete with spares, steel pontoon, motors, generators, and repair-shop fittings, weighs about one hundred tons. Some of the single parts weigh twelve tons; in larger dredges they weigh up to twenty tons. This being so, the necessity of reasonably easy access to the property is apparent. Deposits reached by trail are quite prohibitive; and the dredge sectionalized for mule-

PHOTO No. 5. Portage Creek, on the Little Delta, Alaska.

back transport has yet to be proved a success. Not only do excessive freight rates affect the first cost of a dredge, but they can be felt during its whole operating life. The erection of one of the larger dredges should not be attempted very far from good transportation facilities.

The cost of transporting bulky pieces of machinery is suggested by the following case: From Colon to Puerto Colombia the United Fruit Company used to charge a flat rate of $5 per ton for freight. All pieces weighing over four tons used to pay an excess rate of 150 per cent; over eight tons an excess rate of 200 per cent. A similar excess tariff was charged on the Barranquilla railroad and the river steamers up the Magdelena.

The amount of labor required on a moderate-sized electrically-driven dredge of the California type may be enumerated as follows: One dredge master, three winchmen, six oilers, two mechanics, and several roustabouts. In tropical countries the roustabouts may be natives, but where skilled labor is lacking, the remainder of the crew, with the possible exception of the oilers, must be imported. In addition, a staff for the electric or steam power plant and the shops must be considered. However, for the amount of material handled by a large dredge, its labor cost per yard is low.

Dredges are of different types. Perhaps the oldest form of successful operating dredge is what is known as the New Zealand dredge. It has been greatly used both in its home region and in Australia, where conditions are especially favorable for its use. Due to its lightness of con-

PHOTO No. 6.   Panning on Portage Creek.

struction, it has been extensively used in tropical countries, as well as in Russia and Siberia. Compared with what is known as the California type, the New Zealand type is a light power dredge, usually with a steel pontoon. Its first cost is lower than the California type, and its capacity is smaller. This type of dredge is especially adapted for a small operating company whose acreage or capacity is limited, and whose gravel is moderately light, uncemented, and easily dug. Its chief application lies in those districts where lack of transportation is a salient factor and where heavy pieces of machinery can be handled only at a prohibitive cost. Not only is a considerable saving effected in the first cost, but also in freight charges, cost of erection, and simplicity of repair shops. This saving will offset for some years any higher working cost due to the handling of a smaller yardage than might possibly be dug by the California type.

The California dredge was evolved from the New Zealand in order to meet local conditions, as the latter type proved unequal to the task of handling the heavy ground encountered in California. As a result, modifications of the type were made, and each newly built dredge was altered, and the method of operation changed until a new type of dredge and method of work were evolved. This type of dredge has been enlarged and strengthened and its power tremendously increased.

PHOTO No. 7.   Dredge on Hunker Creek, Klondike Region.

The earlier California dredges were equipped with pan or tray stackers, as are those of the New Zealand type. These have been changed to stackers of the belt-conveyor type or straight flumes. In the latter case, yardage is not so much an object as the recovery of the heavy gold lying mostly on bedrock. Most California dredges are equipped with close-connected bucket lines, and these have been found to dig up greater yardage. The New Zealand dredge digs by means of a head-line alone, but this practice has been found unsatisfactory in California dredging, and all of these now employ spuds in digging. In California most of the hulls were built of wood, but in tropical countries it has been found that steel hulls are more satisfactory.

The amount of gravel handled per month and the operating cost per cubic yard vary within wide limits and depend on the dredge itself, the character of the ground, the size of the gravel, and the depth or height of the bank.

The California dredges vary in size from $3\frac{1}{2}$ cubic feet to 18 cubic feet bucket capacity, and are digging in ground varying in depth from 15 to 90 feet. The field of the modern California dredge lies in

PHOTO No. 8. Thawing Frozen Ground on Hunker Creek,
Yukon Territory.

the heavy cemented gravels, which require ponderous and massive machinery to dig; or in those large areas of deep low-grade gravels lying accessible to cheap transportation facilities. Conditions such as these exist at Marysville and Folsom, California. In Alaska, small heavily-built dredges of from one to three cubic feet bucket capacity, built on California lines, are still working.

Another type of dredge, which is adapted to certain types of ground, is what is known as the Australian pump dredge. This dredge was devised to meet certain conditions existing in New Zealand and Australia; conditions that prohibited the employment of either bucket dredging or ordinary hydraulicking. The most important factor governing the choice of this method is the character of the bedrock.

Where the gold is concentrated or near a bedrock that is rough, hard, and uneven, or where it is deep within the crevices of a shattered or broken bedrock, it is necessary to expose and clean the rock carefully. This, of course, prohibits bucket dredging. If the deposit be too low for hydraulicking, other means must be adopted. It is to meet these conditions that this so-called pump dredge has been devised.

It consists of a set of centrifugal pumps, mounted on pontoons that are floated only when being changed from one working face to another. By its use bedrock is exposed for cleaning. The pump dredge consists of two centrifugal pumps, one known as the pressure pump and the other the gravel pump, mounted, together with motive power, upon a wooden hull or pontoon. While working, the pontoon rests on sills laid upon bedrock. The mining is done by directing a small hydraulic nozzle against the bank. The artificial pressure employed is usually sixty to seventy pounds per square inch, produced by a centrifugal pump of 12 to 14-inch discharge. From a bank the gravel is sluiced through runs, cut in bedrock to the sump, which should never be deeper than 20 feet, which is the limit of the suction. When the bedrock is wavy and dips away from the sump, considerable expense is involved in cutting down the runs to grade; in fact, when the bedrock dips too steeply away from the sump, it sometimes will not pay to move portions of the overlying gravel, which are then left behind. As the gravel is piped, all large boulders are removed and left behind on bedrock.

The gravel from the sump is raised by the gravel pump to the line of sluices, a height which depends upon the depth of the deposit and the fall of the sluice-boxes. A height of over 60 feet requires the installation of a second lift pump. The gravel pump is usually placed side by side with the pressure pump on the barge; but in shallow ground this pump can be worked from the bank and it is then placed on the bedrock. The gravel pump is from 8 to 12-inch discharge. A direct-acting or small centrifugal pump is installed at the head of the sluices to furnish them with wash water. From the ends of the boxes, the tailing flows back into the worked-out portion, and is there retained by brushwood dams.

The total power required depends on the size and capacity of the plant and the height of the lift. It varies from 120 to 260 horsepower. The capacity varies proportionately and ranges from 800 to 2000 yards per day. The time lost in shifting the plant, flooding the workings, building the dams, and in all repairs is a serious item. In some cases it amounts to ten weeks in a season's run of six months. During moving, more men are required than in ordinary running. In the cases quoted, the crews ranged from eleven to fifty men per dredge. The first cost of these dredges is much less than that of a small bucket dredge, which fact renders it more feasible for working small areas of ground.

This method is used effectively where the gravel is deep and no dump is available; or if the deposit be of average thickness—say 25 feet—

with its valuable wash lying between a hard uneven bottom and a tough
overburden full of boulders, this type of dredge is applicable. It must
be borne in mind that a limiting factor is the amount of water in the
ground. If there be too much water, the power required to drain the
ground is excessive; if too little, the time required to fill the pit becomes
too great. Where much buried timber is encountered, this method also
applies. When transport is a serious problem, the lightness of the pump
dredge and its simplicity of installation, compared with that of the
bucket dredge, may render it suitable, in spite of the fact that all other
factors for successful bucket dredging may be present.

While this method of mining has its own application, it is nevertheless
contrary to all economic principles of engineering. Its efficiency is only
30 to 45 per cent, and the amount of water lifted per cubic yard of solid

PHOTO No. 9.   Typical Alaskan Gravels near Circle City.

is enormous, as 95 per cent of all material lifted is water. This moving
of useless material affects the cost, which is apparent in the figures for
the work done during 1907 in Victoria. Over ten million cubic yards
was treated by bucket dredges, and nearly the same amount by pump
dredges. The cost per cubic yard in the former case was under 6 cents
whereas in the latter it was over 20 cents. As, however, the pump
dredge treated all the tough, irregular, and deeper ground, which the
bucket dredge could not handle, this comparison of costs is unfair to
the pump dredge. Conditions being equal, however, the cost of pump
dredging would be from two to two and one-half times greater than
bucket dredging.

So much has been written, and ably written, on bucket dredging in
California that it is useless in a chapter of this scope to go into further
detail upon the subject. The reader is referred to the bibliography at
the end of this chapter, notably the articles by Robert Cranston in the

'Mining and Scientific Press' and the 'Engineering and Mining Journal'; also by John Power Hutchins and various equally notable authorities; as well as Bulletin No. 57 of the publications of the California State Mining Bureau, and the volume of D'Arcy Weatherbe upon gold dredging in California; and the bulletin of the United States Bureau of Mines recently completed by Mr. Charles Janin.

HYDRAULIC MINING.

The term 'hydraulic mining' is here defined as that class of mining employing the use of water under natural pressure in giants or monitors for the pupose of eroding a gravel bank, washing it through sluices, and disposing of the tailing by gravity. Next to the question of gold content, the determining factors in the employment of this type of mining are the presence of an abundance of cheap water that can be brought to the mine under pressure and the existence of sufficient grade for the disposal of the tailings. This latter feature, known among miners as the 'dump' of a property, is exceedingly important.

The size of the gravel is not so important in hydraulic mining as in dredging. Much larger boulders can be handled, provided there is room for their disposal. Fairly tight layers of clay and cemented gravel can be first shattered and then disintegrated by the giant. Gravels of 600 feet or even greater depth have been worked. In fact, with the use of a benching system, there is practically no limit to the depth of the bank that can be exploited. Of course, other things being equal, the smaller and looser the gravel, the higher will be the duty of the water.

The influence of bedrock is not so important in hydraulic mining as it is in dredging, yet an ideal bedrock is one that is even, does not disintegrate readily, and is soft enough not to cause too much expense in cutting sluice-ways, and is still hard enough for easy cleaning. It is a great help if the natural grade of the channel assists in setting sluiceways. A deeply shattered bedrock is difficult and expensive to clean, as the gold finds its way deep down into the crevices.

In this connection, a type of granite bedrock encountered by the writer in tropical countries and other places where the rainfall is heavy, may be mentioned. It weathers underneath the gravel to depths of 10 feet and over and forms a soft hummocky under-burden to the deposit, all of which, though absolutely barren, must be cut and washed away in water to secure a recovery of the bedrock gold, lying for the most part on top of it and at the bottom of the gravel.

The existence of an ample and cheap water supply is imperative for this class of mining. On the question of water depends in great measure, that of the duty and subsequent operating cost. The character of the ground through which the ditch is to be run should be carefully considered as to seepages, slides, footings for flume posts, cross ravines, general liability to washouts, and its ability to withstand erosion.

Effective heads range generally from 200 to 600 feet. Below the former figure the duty is apt to be low. Above the latter the requirements of extra heavy pipe, anchorage, and bracing, become exacting. Within limits the amount of water required varies inversely with the head obtainable. With a 200-foot head a flow of 2000 inches would suffice to operate a moderately sized hydraulic mine, while with a 400-

foot head 1000 inches would move nearly the same amount of gravel, provided it were loose and free.

Few hydraulic mines are so situated that a full head of water is obtainable throughout the year, consequently part of the excess of water during the season of greatest flow is conserved wherever possible, in storage reservoirs. When the supply becomes low, these are frequently allowed to fill during twelve to twenty hours of the day, the water being used for only a short interval.

Where it is impossible, from the contour of the country, to obtain water under a natural head of any magnitude, attempts have been made to substitute artificial pressure from pumps. This is only practicable where water power is cheap, as well as abundant. The most noteworthy recent instance of large-scale operation of this sort, which has come under the writer's observation, is the hydraulic work which was done on the Panama Canal, where sea water was used under pressure for cutting down the material with giants, centrifugal pumps being employed to elevate and discharge the material. The latter was chiefly silt and fine sand, but even under these circumstancs the duty was low and the cost high. While such installations may prove successful mechanically, they have in most instances been failures financially.

The duty of water, in cubic yards per miner's inch, varies from one to ten, and depends upon the character of the gravel, the facility for disposal of tailings, the amount and head of water available.

The presence of an ample supply of good timber for construction is more vital than in the case of dredging, as it is constantly in demand for the building and repair of flumes. sluices, trestles, dams, sluice linings, giant and pipe line bracing, as well as the manufacture of sluice-blocks. The presence of buried timbers in the deposits is not at all prohibitive in this class of mining, as they can readily be piped out and cut to size convenient for the derrick. The clearing of a dense growth of standing timber, of course, adds to the cost of mining.

After the equipment and installation of a hydraulic property, the question of labor is not a serious one, as but few skilled men are required. Three pipers, six sluice men and ditchtenders, a good blacksmith, and perhaps a winchman are all that are required on a mine of moderate size, handling from two thousand to five thousand yards of gravel a day.

One of the chief advantages that hydraulic mining has over dredging is the fact that it is not so greatly affected by the cost of transport. A 7-foot dredge complete weighs nearly one thousand tons, whereas the equipment of a hydraulic mine, handling the same amount of gravel, complete with giants, pipe, gates, sawmill, derrick, and tools, weighs one hundred tons.

Besides the cost of equipment in the subsequent running of the mine, that of supplies and spares amounts to a comparatively small sum, most of the repairs being made by the blacksmith. For the above reasons, and owing to the fact that all equipment can be shipped in small packages, a hydraulic mine can readily be operated in districts reached only by trail. Where timber must be brought, the question of transport becomes much more serious; so serious, indeed, that in some cases it may prove prohibitive.

The question of the impounding of debris is a serious factor in plans for hydraulic mining in those districts in which debris restrictions are enforced. The problem in California has been fully discussed in another chapter of this volume.

In planning the operation of a property, it is important to consider the effect of the disposal of tailings upon farming and other lands. Where irrigation canals are fed from rivers below the dumping ground of the mine it is quite possible that these canals may be silted by mining operations, which would naturally result in trouble for all concerned.

An ideal dumping ground is one like that of the LaGrange, in northern California. The tailing is dumped into a narrow valley about four miles long, which is the property of the company and drains directly into the Trinity River. This river and the Klamath were not included in the prohibition of the debris law. They are not navigable streams, and the damage done by hydraulic mining to the farming interests is practically negligible.

Where all factors calling for the employment of hydraulic mining are present, with the exception of dumping facility, this adverse condition can sometimes be overcome by the use of the hydraulic elevator .

Cheap and abundant water under great pressure is essential. This is due to the fact that the efficiency of these machines is notoriously low. The lifting power of the elevator is from one-sixth to one-tenth of the head of water employed. The proportion of solids in the total weight of material lifted is at most 5 per cent, the remaining 95 per cent being water. It is largely due to these facts that mining by this method has not always been successful. Another factor requiring consideration is the size and amount of boulders in the ground. Ordinarily, with an ample supply of water, in regular hydraulic mining, the size of the boulders that can be ejected through the sluices is in the main limited only by the width of the latter. In hydraulic elevation, however, the limit is governed by the width of the throat of the elevator; consequently, as this, even in a large elevator, is not more than 15 inches, all boulders of larger diameter must either be blasted or handled by the derrick or stone-boat. Consequently only ground containing moderately small gravel is adaptable to this method of treatment.

When a deposit of gravel, not exceeding 50 feet in depth, lies on a hard bedrock with little or no dumping facility, and the water supply is limited, and when nests of heavy boulders exist, a modification of the hydraulic method with the adoption of an inclined grizzly or of a simple inclined sluice may be advisable. The former, sometimes called the Ruble elevator, will be described in detail later. By the use of it all the heavy tailing is stacked and left on the worked-out bedrock. The grizzly is unsuitable on rough and extremely wavy or uneven bedrock, owing to the difficulty in moving. Its chief advantage over all other forms of elevator lies in the low cost of construction, as well as the fact that it can be made on the ground. Another point in which it excels is its capacity for handling heavy boulders.

The inclined sluice is employed to elevate material that has already passed through a string of sluice boxes; or to elevate material from the diggings to the sluice boxes in order to secure dump. Small deposits in northern California and southern Oregon have been worked by this method and also deposits in the Circle and Forty Mile districts in Alaska. On Mastodon Creek, in the Circle district, a small plant is in operation, which may be cited as an example of this practice. Here two No. 1 giants and one No. 2 giant have been operating with water under a 100-foot head. The sluice-boxes, 30 inches wide, of which there were six, each 12 feet long, delivered their tailings into a common sump. From here the tailing was stacked 35 feet high through an inclined sluice by the No. 2 giant.

The depth of the gravel was 9 feet, and the duty, including the water used for stacking, was $2\frac{1}{2}$ cubic yards per miner's inch of twenty hours.

Hydraulic mining methods vary greatly in different districts according to the physical conditions encountered. Even in California a case may be cited of two totally different practices in regions only a few hundred miles apart. In the northern part of the state in territory tributary to the Klamath and Trinity rivers, the great majority of the bars which have been worked are not very high above the beds of the present rivers. The depths of the gravel banks do not generally exceed from 30 to 50 feet. For this reason, the employment of a drive or 'booster giant' is necessary. This giant is set in such a position as to drive the gravel directly to the sluice boxes after it has been cut down.

In most of the larger mines of the Sierras in eastern California, hydraulic mining practice is totally different. Here the banks are high, running to 600 feet in depth, and as a result, resetting of giants is far less frequent. In the practice in vogue in this territory the giants are employed mainly for the purpose of cutting down and are set directly in front of the bank. For transportation, lead water is relied upon, as the grade and dump are much better than in the northern country, where most of the water is applied through the nozzle and but little lead water is used.

In planning the operation of a hydraulic mine, the first question to arise is that of getting the needful supply of water under working pressure. This is usually the most costly feature of the mine. A ditch and flume several miles in length may have to be dug and constructed. A timber or stone or concrete dam may have to be blasted in the river or stream at the point of diversion of the waters or a tunnel may have to be cut through the solid rock for several hundred feet in order to avoid the washing out of the flume by annual floods characteristic of torrential streams.

In a chapter whose scope is limited, it is impossible to give the more intimate details of this type of work. The reader is referred to the bibliography at the end of this chapter for details of construction; however, the broad general principles of procedure will be briefly outlined.

As a rule, in all temperate countries, the first thing to be constructed is a sawmill. If possible, this is usually located near the head of the ditch, either above or below the point of diversion of the water. As fast as the ditch and flume are built, water can thus be used to carry lumber to the point where it is needed and later on to the mine itself. Most dams in timber countries are built of a crib work or cross timbers laid upon sills, which are placed in hitches cut in the bedrock at a point where a solid foundation can readily be obtained. This crib work is filled in with boulders, rocks, and sand, and in a short time the material brought down by the river will make it a permanent structure. As a rule, it is faced on the upper side with heavy planking. The turnout gate for the flume should be sectionalized in order to control the amount of water admitted to the flume. One very good system is to have the gate made of 6 x 4's, which can be raised independently of one another, thereby widening the aperture into the flume or narrowing it whenever desired with the least expenditure of effort. The first or head boxes of the flume should be protected either by a bedrock wall or a built-up wall against possible flooding and raising of the river.

In the location of the saw mill, advantage is usually taken of some small gully running back up the hill toward the thickest timber for the building of a skid-way, and as fast as the logs are cut the regulation lengths, they are shot down the skid-way into a pond or against an embankment immediately above the mill. If the mill is at or near the ditch line, where the latter has already gained enough distance above the river, the ditch will supply power for running the sawmill. When a considerable supply of timber and lumber has thus been assured for flume and dam purposes, a great deal of judgment must be used not only in the flume construction, but in digging the ditch. In flume construction, the question of footings is most important and will often determine the life of the flume.

With regard to the general features of the ditch, it may be said that the grade should be suitable for the quality of the soil, the front bank should be wide and firm, and the hill bank should be well sloped. The main idea of its design is to keep the force of the water, as well as the greater portion of the friction back against the hill as much as possible, yet not so much as to cause extensive undercutting and slides. To avoid these latter nightmares of the hydraulic miner, even when the ditch has been well constructed, one or more ditchtenders may be necessary. There are, of course, innumerable methods of ditch repairing and bracing known to the experienced miner. As it is impossible in limited space to give too much of detail, the reader must again be referred to the bibliography.

Turnouts and sand gates at frequent intervals are essential, and in case of breaks in the ditch, may save much damage. Changes of grade should be avoided unless there is marked change in the character of the ground, as the point of change merely fills to the regular grade with cuttings brought down by the water. If the grade used for the roughest surface be carried throughout the ditch, things are on the safe side. The flume grade will of course be constant and much less than the ditch grade. Care should be used in blasting through hard rock to prevent excessive fracture and consequent leakage. The whole ditch should be carefully puddled with clay or saw-dust at frequent intervals.

Ground that is apt to slide should be flumed in spite of the temptation to dig the ditch in soft ground. It will eventually have to be flumed anyway, and this, when the footing has slid away, will be done at a much greater cost. Above all, it should be borne in mind while constructing a ditch and at all other times that when the giants are not working there is no money going into the sluices and nothing is more costly than shutting off water to repair ditch breaks during the mining season.

Having brought the water above the property with the desired head, the next thing requiring careful work is to get it down to the working face. If the pressure is great and the hillside steep, the greatest care must be taken in bracing the lower portion of the pipe line. The details of line, gate, and elbow bracing, as well as penstock building, are matters of practical importance, and the reader is again referred to the bibliography for details of construction.

When everything is ready to start laying the branch strings of pipes and setting the giants, the question of the proper method of working the ground arises. This calls for the exercise of good judgment. Ground that has only an ordinary dump into a river that depends upon annual floods to carry away excess tailing should never be opened from the lower end, as a dam will be formed, and all detritus from the workings, including the fines, will fill up the river and spoil what dump there is. On the other hand, in case of good dumping facilities, it is well to open at the lower end and take advantage of the bedrock slope in laying sluices. If two strings of sluices can be so laid that one giant can be set between and turned from one to the other while heavy boulders are being blasted or removed by derrick, efficiency is gained. In case a dump giant has to be used to pipe dump up river—it should never be piped down—the water may be turned into it at similar odd times. The use and placing of a derrick, equipped with chains, for the larger boulders, and stone-boats for the smaller ones, requires judgment and experience. Care must always be taken not to cause blocks with boulder piles.

As stated before, there are two general types of California practice in hydraulic mining methods brought about by different sets of conditions. In the eastern portion of the state, tributary to the Sacramento River, and now practically closed by the debris law, gravel banks are for the most part, deep and have excellent dump. As a result, when bedrock cuts were started for sluice-ways, their depth was usually so calculated that when a string of sluices reached the limit of the workable ground, it would still be in bedrock. Thus the sluices were always kept up to the face, and all giants were employed in cutting down the bank alone. This is very different from northern California practice, in which cutting is done by the field giant, which is, in shallow ground, kept close to bedrock. Driving across bedrock to the head of the boxes is done by the booster giant, which is often placed on the bank in line with the boxes. By the shallow cut thus necessary for the sluice-way many feet of dump are saved.

Speed and efficiency in handling ground are largely dependent on the personal factors of judgment and experience. Gold saving, on the other hand, is more a matter of common sense. On this point very much has been written, and the reader is again referred to the bibliography at the close of this chapter. A general discussion of the fundamental principles will, however, be confined to as short a space as possible.

For the saving of fine gold, it is generally conceded that railway iron (usually about forty pounds weight per yard), laid across the boxes, and spaced at 2 to 2½ inches apart, or Hungarian riffles, which are 2 x 4 scantlings covered with steel straps, are the best form of riffles both for security and durability. These riffles are often placed in sections, which alternately run lengthwise and then crosswise of the boxes. In many places it is not always possible to secure the necessary material, and sluice blocks sawed from fir trees, about ten inches thick, are perhaps more generally used than anything else. These sluice boxes are separated from one another by pebbles placed between them in order to form riffles two or three inches in width and the depth of the block. Details of construction of sluices and of the making of riffles can be found in any of numerous reference books.

In the case of fine gold, one or more undercurrents should always be used, no matter how long may be the string of sluice boxes. Quicksilver is usually left out of the first few head boxes, but in the lower boxes should be used carefully, too much being nearly as harmful as

PHOTO No. 10.   Hydraulic Mining near Forks of Salmon, Siskiyou County, Cal.

too little. Care, of course, must always be used in planing and caulking sluice bottoms to avoid loss. Flooding of sluices by the pipes, with its constant fluctuation, should never be allowed. To avoid this, where possible, some lead water should be employed independent of the giant water. The grade of sluices depends upon local conditions, varying generally from 3 to 9 inches per 12-foot box; 6 and 7 inches per box are good grades with which to operate.

It has already been mentioned that in deposits without sufficient dump for ordinary hydraulicking, elevators may be used. There are many deposits in California and elsewhere that have been successfully worked by this means, thanks to the presence of abundant cheap water.

The usual means of operation is to blast out a pit or sump about 4 feet deep and 10 or 15 feet square in the bedrock. The receiving end of the elevator is set in the lowest portion of the pit. The sump should be centrally placed, as a great deal of time is required to move the elevator, with consequent loss of water season. Except where the gravel is very fine, a derrick and stone-boat are used in connection. A grizzly is placed in front of the intake, whose bars are spaced at least one inch closer together than the width of the elevator throat. At frequent intervals the grizzly must be cleared by hand labor, assisted by the stone-boat. As a result, during these intervals there is a loss or waste of power which varies with the amount of boulders in the ground.

The efficiency of these machines is low for the amount of water consumed. In the writer's experience of the principal types now on the market, the Evans, Campbell, and the Hendy elevators are probably the most efficient. The Evans elevator has a distinct advantage, due to the air suctions used on either side, which prevents the formation of a

PHOTO No. 12.   Ruble Elevator at Gilta, California.

vacuum in the discharge column. It is certainly true that this machine will give a greater lift for a smaller head than any other type that the writer has used. It has, however, some disadvantages on account of its lightness of construction.

One of the most successful elevators in northern California was constructed on the mine at the property of the North Fork Salmon River Mining Company in California. It consisted of a steel-lined box about 18 inches square inclined at an angle of about 50 degrees, with a 5-inch jet at the bottom and a half-inch steel striking-plate at the top, immediately over the sluice. The material was elevated about 30 feet. The operating head was about 250 feet. The water supply was exceptionally good and about 700 inches was required for the elevator. Boulders up to 11 or 12-inch diameter were handled, and there was no attempt at constructing a throat, everything being lifted directly on the jet. The capacity was approximately 1000 yards per day.

As before stated, the Ruble elevator is simple and is usually constructed on the ground where it is to be operated. It consists of an inclined chute at an angle of about 17 degrees and about 100 feet in length, including a 10-foot apron, which is used to make connection with bedrock. This apron is set with its walls fitting closely inside the walls of the main elevator chute, which is about 90 feet long. When moving, the apron is moved separately.

The chute itself, as well as the apron, is lined with quarter-inch steel sides and three-eighths-inch plates on the bottom. It is about 8 feet in width, and the walls taper from 12 feet in height at the bottom to about 4 feet at the top. For the first 20 feet of the incline the bottom is solid.

PHOTO No. 13. 'Boiling out' with Ruble Elevator.

The remaining 70 feet consists of grizzly-bars spaced about 2½ inches apart, running transversely; these bars are made of 2 by 6-inch timbers covered with straps of half-inch steel. Underneath the grizzly is a false bottom which slopes down from the upper end of the elevator to the sluice-box, which in turn runs out at right angles from under the main chute, directly under the lower end of the grizzly. Both the false bottom and the box are lined with light steel, and 6-foot extensions are built from the latter on separate bents, set at the desired grade, until about 60 feet of sluice is obtained. The rifles used are Hungarian, 2 by 4-inch timbers covered with strap steel, set at intervals lengthwise, but mainly crosswise, of the sluice. If desired, the sluice can be covered and put under lock and key.

The supporting structure of the main elevator is built on three long stringers about 12 by 14 inches in size. Bents 4 by 6 inches are built crosswise upon them to support the chute. Iron rods may be used for bracing. The whole grizzly is mounted upon skids set crossways of the stringers and supported by blocks resting upon bedrock. When moving, a roadway for these skids has to be built ahead of it. Moving is done with a light winch and a cable, one mule furnishing ample power.

The method of operation is as follows: The elevator giant is set squarely in front and in line with the middle of the grizzly, about 80 feet back from the bottom. Wings are built out on either side of the elevator, one wing reaching clear to the bank. The wings are about 10 feet high, and faced with the poorest timber scrap on the place as a protection. The supporting frames can be made so as to be portable and quickly set up by bolts. They must be well braced from behind, as gravel is constantly being slammed against them by the giant.

The field or cutting giant starts operations behind the elevator giant and works along the face from bedrock, taking a layer from 10 to 20 feet thick, driving it up along the bank to the wing of the elevator, and piling it. From this point onward it is taken in charge by the elevator piper. As both giants are often in operation at the same time, a rough shed is built over the elevator giant to protect the piper.

The piper must use great care in order to avoid loss of fine gold. He picks up a few cubic yards of material and drives it up the solid portion of the incline. Then he carefully 'boils-out' the fines over the lower portion of the grizzly. The fines go through into the box, and not until the boulders and heavier stones are clean and bare are they pushed over the end of the machine. A few are always left on the grizzly to act as a baffle for flying fines and gold, as the latter has a tendency to flick over the end on the giant spray. After the fine material has thus been separated and put through the grizzly and the boulders driven over the end, the action is repeated.

About every hour or so the giant must be swung to one side, and the fine tailing that is collected and heaped at the discharge end of the sluice piped away. A dump giant is kept at one side to pile up at odd moments when the water may not be in use in either of the other two.

When the field giant has been advanced along the face until it has cut its entire swath, it is drawn back again and another layer of 10 or 20 feet removed. From one setting of the machine a tremendous amount of material can be reached, especially if the pressure is good.

When the boulder dump is filled to the top of the machine, sills are laid on it with a platform of boards as an extension of the elevator. This operation is repeated every two or three days until the dump is piled much higher than the machine. Then the dump giant is run forward, the whole pile piped down in a few hours, and a new start is made.

When the driving limit of the field giant has been reached, the machine must be moved. The time required to do this (5 or 6 days) is one of its chief drawbacks. This can only be obviated by using two or three machines and changing the water from one to another. In this way, two machines can always be kept in operation while one is being moved. With a 50-foot bank, and the field giant working under 400 feet of pressure, the elevator should be good for at least four weeks work in one place.

As previously stated, the machine is moved by winch and cable. Great care must therefore be taken to have the skids level and firm, so as not to rack it.

The writer has operated a machine of this type on ground carrying about 10 cents per cubic yard, with about 8 feet of dump into a small river. From 600 to 1200 inches of water under about 450 feet head was available for about four months of the year. During this time 100,000 yards of gravel was handled, and the total operating expenses, including ditch maintenance and all preparatory work, was $6,000, or 6 cents a yard. As the gravel was heavy, containing nests of boulders from 1 to 5 tons in weight near bedrock, and the bank was only 20 to 25 feet in height, this may be considered a fairly low working cost. The powder bill for the season, with powder at 25 cents a pound, was about $200,

PHOTO No. 14. Showing the Size of Boulders put through the Ruble Elevator.

because all rocks that would not conveniently go over the grizzly had to be blasted to about half-ton size, or a trifle less. Boulders weighing more than a ton could be put over the machine; although of course very slowly, so that it was not good practice to do it.

Although the mine was distant 90 miles from the railway and all steel had to be imported, the total cost of the machine erected upon the ground was about $3,500. The capacity was from 1000 to 2000 yards per day of 24 hours. This capacity could be increased at least 50 and perhaps 100 per cent by the use of an automatic gate at the upper end of the machine, which would keep all fine gold from flying over, and obviate the necessity of losing time in careful boiling-out of the fines. This would also prevent any loss due to careless piping, as the material could be jammed through the machine almost as fast as the pipe could carry it without danger of gold flying over.

The subject of hydraulic mining, if covered in detail, would occupy a volume in itself. In the foregoing paragraphs, the writer has endeavored to restrain himself to an outline of the fundamental principles that govern the selection and operation of a hydraulic mine. For more complete detail, expressed in a better manner by a man who is a master of his subject, the reader is referred to Mr. A. J. Bowie's treatise on hydraulic mining.

One subject, however, should perhaps be mentioned in addition. The above methods of mining secure their greatest efficiency in gravel that is moderately loose and free. In a great many of the larger deposits of gravel now remaining in California, the material is so tightly bound together that it becomes necessary to blast the banks before hydraulicking.

The usual procedure in this case is to run a tunnel directly into the bank, either on bedrock or near the bottom of the ground which is being benched off to a distance which will approximately equal the height of the bench. From the end of this tunnel cross cuts are run at right angles in directions so that the whole working resembles a 'T' in form. This tunnel is carefully packed with explosives according to the mass of the ground which is to be broken, and the hole is then carefully sealed and tamped, connection being first made for electric detonation. After the bank has been shattered and broken by this discharge, it is eroded and broken down by the hydraulic giant in the usual manner. The cost of this blasting will, of course, vary with the physical conditions which obtain in the bank. As a rule, it will vary between 2 to 5 cents a yard.

DRIFT MINING.

Gravel that is covered by flows of igneous rock or by a heavy deposit of overburden, and the metal in which is concentrated within a relatively narrow strata, is usually mined by drifting.

The relative importance of the factors that govern the choice of this form of mining is often dependent on the conditions under which operation must be undertaken. For instance, in the early days of California mining, many deposits were drifted for their richer streaks by miners with little or no capital. Later, some of these deposits were hydraulicked with great success. In the same way a large proportion of the Oroville ground was drifted before the advent of the dredge.

The proper conditions for the operation of a drift mine may be stated as follows: Values heavily concentrated in the gravel, without too large boulders, too much water or running sandy ground; bedrock that should not be swelling, and yet which is soft enough to have caught and retained a fair proportion of the bedrock pay or heavy gold. An ideal drifting ground is one in which gravel is about five feet thick on a slate bedrock and is capped by a smooth homogeneous body of lava or volcanic mud. The gravel should be fairly loose, without cement.

In a drift mine skilled miners and timbermen are a necessary adjunct. It is often necessary in opening up such a mine to run long cross-cut tunnels or to sink fairly deep shafts. These add greatly to the cost of opening a mine. A heavy flow of water, which necessitates much pumping, is also a source of expense. For this reason, ground should be opened up by tunnels which are deep enough to drain the channel if possible. Running and sandy ground requires closer timbering and extreme vigilance. Nests of large boulders bring up the cost of the mining, as they must be blasted.

Where the deposit is buried under a heavy lava flow, drift mining is imperative. Where the rich streak is buried under a deep overburden of barren material this method is often advisable. In the case of frozen ground, as in Alaska, drift mining has been almost universally employed in the smaller holdings. During the winter the gravel is taken out and piled on the dump to be sluiced later with the spring thaw. Steam batteries with points are employed, and the necessity of timbering is obviated by the frozen nature of the ground.

In the following, an abstract has been made of a report on 'Drift Mining in California', by Russell L. Dunn, published in the Eighth Annual Report of the State Mineralogist of California. In part quotations have been made directly.

Drift mining is peculiarly a California development, originating from the conditions of location of these deposits. The earlier channels now cut by the modern streams are usually accessible to bedrock tunnels. The ancient river system, whose buried channels are auriferous, extended from what is now Butte and Plumas counties on the north to Tuolumne on the south, and from the eastern edge of the Sacramento Valley almost to the summit of the sierras. The topography of the country during their period of formation can not now be restored with more than probable certainty. Apparently the ancient river system was similar to the present one in relative location and direction of flow of the main streams.

The ancient streams, judging by the masses of gravel in their channels probably carried larger volumes of water than the present streams, and the mean gradient of their beds was considerably more than that of the existing streams at corresponding points, it being almost certain that the elevation of the Sierra to its present condition and altitude was before the cretaceous period. The general surface of the country was not as rugged as now, being hilly rather than mountainous. The gold in the channels is a product of the primary disintegration of the auriferous slates, talcose rocks, and quartz veins. The erosive agencies of water and cold were probably more powerful then than now. Le Conte says that a period of glacial erosion was prior to the formation of the channels, and was the greater disintegrating force.

The changes in the location of the channels have been made by eruptive agencies and their filling up with accumulations of gravel, sand, and clay. This covering up and obliteration of the surface was not the result of one season of eruptive activity, but several, separated by enormous intervals of time. The first flows probably did not completely divert the streams, except at a few points, but merely raised their beds and changed the character of the channel deposits. The period of inactivity was in time followed by another period of eruption, and in its turn by a period of quiescence. This sequence repeated several times, but with a diminishing power, and finally ended in the complete cessation of the eruptive energy. These latter flows consisted largely of volcanic ash and volcanic mud. The channels and surface depressions generally, and some of the lower hill elevations, became more and more obliterated until at the end of the last period of eruption a completely new topography was forming, the beginning of the present.

The lessening area to the south covered by the successive flows, accounts for the greater erosion of the eruptive deposits in the southern portion, and for the greater depth and more numerous strata of the northern portion. It is probable that many of the existing river channels are from original ones cut deeper into the country rock, the volcanic flows not obliterating them at all. This is particularly the case in the lower courses of the larger streams.

"The old river channels now are—as the result of the eruptive flows first filling, then denudation by glacial and stream erosion—depressions in the surface of the country rock filled with river sands, gravels, and clays, and capped with lava, volcanic ashes, and tufa, with possibly wash gravels lying between the volcanic flows—the remains of stream erosion in the interval between the flows. The depth of the gravels on the bedrock will vary between limits of nothing to 300 feet; the depth of the volcanic flows and other gravel deposits from nothing to 1500 feet; though at no two points would exactly the same deposits, either in quality or relation, be found. The following data from the shaft of the Gray Eagle Drift Mine, Sec. 6, T. 13 N., R. 10 E., M. D. M., near Forest Hill, Placer County, is typical, and well illustrates the phenomena of several of the eruptive periods and the stream flows of the intervals between. Beginning at the surface, in sinking, the shaft passed through—

Red soil and loam	10 feet
Soft gray volcanic ash	31 feet
Hard gray lava, containing angular fragments of slate	80 feet
River wash, sand and gravel in alternate strata, principally sand	34 feet
River wash, gravel and sand in alternate strata, principally gravel	30 feet
Yellow water sediment, pipe clay	25 feet
Loam, fine black sediment, containing leaves, logs, etc.	10 feet
Large boulders, water worn	10 feet
Hard, chocolate-colored lava	60 feet
River wash, gravel and sand	10 feet
Hard, chocolate-colored lava, containing logs, some petrified	20 feet
River wash gravel	7 feet
Hard, chocolate-colored lava	25 feet

"At this point the country rock is struck sloping down, showing that the bottom of the channel has not been reached. On and in this rock, gold was found.

"In this particular case there are four distinct lava flows determinable and four river flows in substantially the same channel. Not till the channel became full by the last volcanic flow did the old stream take an entirely different location. Comparatively few shafts have been sunk through these lava flows, the mining of the auriferous gravels underneath being most practicable through tunnels, and in the sinking of the shafts but little attention has been paid to keeping a record of the character of the ground passed through. However, in the workings of some of the drift mines through tunnels, several of these lava flows have been located far underground, not superimposed one on the other, but filling channels that have been cut through and crossed older channels filled with older lava flows. In the Bald Mountain Mine, at Forest City, Sierra County, the channel being mined was crossed and cut through by another channel about 500 feet wide. This latter was filled at the bottom with a kind of volcanic mud and contained no gold. In the Mountain Gate Mine, at Damascus, Placer County, a wide white quartz channel was found to be cut through and crossed by another channel over 500 feet wide and 60 feet lower at the crossing. This last channel, unlike that in the Bald Mountain Mine, contained auriferous blue gravel (almost exclusively slate) from 6 to 15 feet in depth, directly overlaid with a hard, compact lava. In the Paragon Mine, at Bath, Placer County, there are three distinct determinable channels. First, the lowest and original, a blue gravel channel lying directly on the country rock. Second, an upper channel 150 feet above the first in an elevation and having the same general line of flow. Between the two are alternate layers of wash gravel, sand, and pipe clay. Third, a channel crossing and cutting through the second, but not down to the first. This last is filled with a lava flow.

"Some of these old river channels are filled to depths of several hundred feet with gravel, sand, and pipe clay, all river deposits, which extend to great widths and far beyond the limits of the lowest channel depression. These immense accumulations of gravel and other detrital matter, in a less degree than the eruptive flows, have still been the causes of changes in the location of the channels. An example of this kind of change, which is more than usually marked, exists in the channels in the vicinity of Forest Hill, Placer County. Four miles above Forest Hill there is only one channel traceable by surface indications; a mile nearer Forest Hill it seems to have had two distinct beds and locations. One of these runs south through the Paragon Mine, in which it has been followed for almost 8000 feet, thence cut off and eroded away for over a mile by Volcano Cañon, it reappears as the extremely rich front channel of Forest Hill, having there a southwesterly course. The other, first having a southwesterly course till it is a mile west of the Paragon channel, then turns south, running through the Mayflower Mine, in which it has been followed for about 2500 feet, and keeping the same general direction it finally joins the other about a mile and a half southwest of Forest Hill. It seems almost impossible that both of these should have been made at the same time, but they are undoubtedly the work of the same stream, though the points of parting and reuniting have as yet not been found. Their common origin shows itself in the similar character of the gravel wash in both, and the similar character and yield of gold; also the widths of the beds of the channels are practically the same, and the elevations of corresponding points in these beds in agreement. The extreme rise of the surface of the country rock between the two channel beds, so far as known, except at a few points of no extent, does not appear to have exceeded 150 feet.

"The theory (an opinion) of these two channels is, that the first cut out by the stream became, in the end, filled with gravels and other water deposits until the water flow was forced over the low elevation between into the channel of a tributary, which it cut out and made into the main channel till it in turn became filled with gravels and detrital matter up to the level of the other. From this time on, the location of the channel was probably not permanently fixed, as wash gravel of similar character is deposited all over the country rock between the two channels, and all contains some gold. What has already been noted as the second, or upper channel, in the Paragon Mine was, from its unusal richness in gold for gravel so far above the surface of the country rock, the probable location of the flow for a long period of time. Both channels, the country rock and overlying gravels between, are covered with 200 feet depth of lava, on which is another deposit of wash gravel from 20 to 50 feet in depth, containing some gold, and over this a second lava and volcanic ash flow capped with the surface soil, from 100 to 300 feet in depth."

The filling in and covering of the old channel deposits was not uniform nor was the subsequent denudation. The portions of the channels, in which were the largest accumulation, seem to have been lightly covered and subsequently eroded so that the remains of these larger deposits, where they have not been obliterated, are now in the form of gravel hills, being the summits of the ridges between the present river cañons.

The early miners worked the more shallow of these 'hill diggings', and discovered that the richest gold-bearing gravels lay immediately on the country rock and followed it into the mountainside. This branch of the gold mining industry soon became known as drift mining.

Before discussing the details of methods and appliances, the conditions of drift mining are expressed as follows: The auriferous placer deposit is river-washed gravel, most often lying in a narrow depression of the surface of the country rock, overlaid with either comparatively barren gravel and the detritus of fresh water erosion from a few feet to several hundred in depth or with lava and volcanic flows to as much as a thousand feet in depth, or with both, in varying relative proportions and alternation, depending on the surface denudation during the period of intermittent eruptive activity and since its close. Reliable surface indications of these ancient channel depressions are practically limited to the places where they are uncovered by erosion or cut off or into by the present precipitous stream cañons. The present main river cañons have cut down hundreds of feet lower than these old channels in all but a few localities.

Experience has shown that of all the old channels, those that are the oldest and that are invariably on the bedrock must surely contain gold in sufficient amount to justify prospecting and working. Top wash channels, or sometimes a stratum of gravel in the channel many feet above the bedrock, are found to contain sufficient gold to make drifting profitable, but such instances are not common. Not all of the oldest channels contain pay leads, although they almost invariably contain some gold. The pay lead in these channels is often an uncertain quantity. It takes its own course between the rims, and sometimes on them. The pay lead is usually close to bedrock, and if the latter is soft or creviced, it is in it. Sometimes the pay lead is the full width of the channel. More often it is only a comparatively narrow line, meandering through it, first abutting on one rim, then on the other. It is not always continuous, being broken by barren places. Great variations in gold yield will occur in the same lead, due partly to the currents of the old stream and partly to the fact that the heavy gold has not been moved very far from the location of its original matrix. Occasionally, large bodies of pay are found on the rims at a considerable distance from the bottom of the channel. These were probably the deposits on the old channel beds which have been left above as the cut became deeper. The nature of the bedrock bears some relation to the gold distribution. A soft slate is favorable, the gold being found in it to the depth of a foot or more. A rock full of seams and crevices and with a slightly irregular surface is better than a hard, smooth, water-worn surface.

In prospecting for a channel, if an inlet, or an outlet, or a break can be found, the location of the channel is not difficult. The principal difficulty to be contended with is the determining of the rise of the rim above the channel bed. A cross rim at an outlet or an inlet often shows by the richness of the gravel on it that it was originally part of the bed of the channel.

Whether the point of opening is at an outlet or inlet necessarily has considerable to do with both the prospecting and final opening of the mine. An outlet is naturally the most favorable point from which to open a drift mine. Prospecting from an outlet requires much less work to obtain the necessary information on which to open the mine than where the prospecting is from an inlet. The proper method of prospecting is to run a tunnel through the rim as near as can be determined in the direction of the channel and far enough down in the rim so that when it breaks through the bedrock it will be in the bed of the channel. When prospecting at an inlet, the tunnel should be some distance under the bed. There is always a possibility that the prospect tunnel may be lower than the channel or it may miss in direction and run off to one side in the rim of the channel. To guard against possible mistakes, it is advisable to make an upraise from the tunnel as soon as it is believed to be through the rim.

The development being at an outlet, the channel will be prospected and worked into on the ascending grade, assuring the most economical and perfect drainage At an inlet the prospect tunnel must be run relatively lower, and therefore longer, in order to gain on the descending grade of the channel The channel once found, it must be followed further and its grade determined with all possible accuracy before opening the mine for work. In the case of a breakout, probably the surest method of prospecting is to either run a slope on the pitch of the rim or to sink a vertical shaft on the presumed line of the channel. When the bed of the channel is located, it is prospected by cross and lateral drifts to ascertain the width, direction, and grade; and the location, extent, and character of the pay lead.

The principal difficulty to be contended with in prospecting is the drainage of the underground water. If the tunnel be too high, and it is necessary to sink shafts or inclines from it to the channel, a flow of water, which may greatly increase the cost of handling material and which may even render it impossible to get to the bottom of the channel, is apt to be encountered. It is then necessary to run a lower tunnel to secure drainage.

All channels have not outlets, inlets, or breakouts that can be found and identified as such. In the case of a channel in which there is only a thin body of gravel covered by many hundred feet of lava, the discovery of exposed gravel is a very difficult matter. Often it may be necessary to spend much money in prospecting to determine the location of such a channel.

It is only since the origin and character of the gold-bearing deposits have been understood that engineering science to locate and develop them has been available. At first its application was quite simple in character, consisting of obtaining the grade and direction of a channel deposit already developed. By this means a point of probable location in its projection was prospected by means of a tunnel or shaft. The Hidden Treasure Mine in Placer County was determined in this manner.

Ordinarily, the location of a channel, in the absence of surface indications, is a complex problem involving a survey and engineering examination of a large territory. The initial step is an accurate topographical survey of the country. This consists, first, of transit lines following the line of contact between the rim rock and the lava or gravel, as the case may be. If this can be made continuous all around the presumed channel so much the better. If, as is usually the case, this would include a greater area of country than is necessary, the transit lines follow the rim lines on both sides of the ridge and are connected by cross transit lines; also connected with these are transit surveys of the underground works in adjacent property. All of these lines are leveled and a plat is made.

The problem of the location of the channel then resolves itself to the determination of the lowest point of depression between the two rims. If the distance apart of the rim lines is not too great, the prospecting problem is simplified. If it is, the problem is complicated on account of the possibility of there being two or three channels with bedrock ridges between.

"In projecting trial locations for prospecting work, the work in adjacent mines is of great assistance because some fair degree of accuracy can be used in determining the rate of pitch of the rims, the grade of the channel, its relative elevation and approximate direction. With these data trial cross-sections from rim to rim at several points can be projected on the plat. If the adjacent workings do not give data that can be utilized for this purpose, it becomes necessary to assume a mean rate of pitch for the rims and a grade for the channels. With these, trial cross-sections are constructed as before, possibly several trial values being tested, till the cross-sections locate the several possible points of channel depression with a fair degree of possible relation between themselves. Formulating the method of determining the channel point in a cross-section line, the rate of pitch of both rims being assumed the same, it is as follows: First, the horizontal distance from the surface point of either rim to the channel point is one-half of the horizontal distance from the first point to the point of intersection of the horizontal distance line lying in the plane of the cross-section, with the bedrock slope line, projected if necessary, of the other rim in the same plane. Second, difference of elevation between either rim point and the channel point is the horizontal distance first obtained for the same rim point multiplied by the tangent of the angle of pitch of rim. From the plat the distance between the two rim points can be scaled on the cross-section line, also the difference of elevation between the same two points can be taken from the contours, interpolations being made if needed, as it is not necessary to place them closer than twenty feet apart.

"Except in probably wide channels (two hundred feet and upwards) no attention need be paid to the width of the channel bed in determining the elevation of channel points from the trial cross-sections. In practice the graphical method of obtaining these points is sufficiently accurate, and has the advantage of rapidity. With the trial channel points platted, a profile of the presumed bed of the channel can be made, using the absolute elevations obtained from the rim points, and its probability determined. Possibly, several sets of projections and profiles have to be made before the one of the greatest probability is determined on. From the platted line of the presumed location of the channel and the surface contours, the most available line for a tunnel, or the best point from which to sink a shaft, can be readily located; also the approximate length of the one or the depth of the other can be determined. The subsequent location of the tunnel line or shaft point on the ground is a simple matter. Except under exceptional conditions, the tunnel line is always selected, the point of entrance on the surface being so placed that when the tunnel has been run with a light ascending grade to the presumed channel line, it will be from twenty to forty feet below the bed of the channel. In order to avoid being too high with the tunnel, it is advisable to run it so as to come at least twenty feet lower than the trial location of the bed of the channel. This additional depth is utilizable for working down stream. To give a greater assurance of certainty as to the correctness of the location of the channel, where the construction of long and costly tunnels is necessary, it is recommended that bore holes be drilled to the bedrock on the presumed line of the channel, and on a cross-section line close to where the tunnel line is located. Their depth being known, the shape and elevation of the bedrock can be compared with the approximations and estimates of the surveys and plat, and inaccuracies in the location of the channel corrected. After the mine is opened, the bore holes can be used for ventilation. Independent of any elaborate topographical surveying (as the writer is advised), a number of such holes were drilled on a mining claim near Gibsonville, in Sierra County, and the location of the channel line determined. Subsequent development by a tunnel verified the correctness of the location.

"The writer has employed this method of engineering determination successfully in a number of instances. It was also applied by Ross E. Brown, E.M., to locate the up-stream portion of the cross blue-lead channel first discovered in the Mountain Gate Mine at Damascus, and already referred to in this article. The discovery point was a mile and a quarter underground, and so situated that, though its course, elevation, and grade were determined, the pitch of the rims could not be. A mile to the northeast the north line of contact between the bedrock and cement was picked up, traced, and surveyed for eight miles up the ridge known as the Forks House Divide. The corresponding south line of contact was traced and surveyed the same distance, the two lines being about 8000 feet apart. On the south line traces of small channels and one important one in the Dam claim (probably an inlet to this channel) opened by a tunnel for several thousand feet, all being evidently inlets of tributaries to the Mountain Gate channel, were found. No main inlet of the main channel on either side of the divide was discovered, but the survey connections made from the underground discovery point in the Mountain Gate Mine with the underground works of the Golden Fleece Mine, five miles to the northeast, indicated that the channel in the latter was the continuation of that in the former. The problem was to locate the line of channel in the intervening country between these two points. These underground works being platted, the approximate distance by the presumed channel line between the two points was obtained, and this, with the difference of elevation, gave a mean trial grade from which the approximate elevation of the intermediate points could be determined. The location of these points between the rims was determined by assuming a trial degree of pitch for the bedrock of both rims, and locating them accordingly. A check on the value given to the pitch of the rims was had in the comparison of the figures of elevation of the bed of the channel in the same cross-section obtained, respectively, from the mean trial grade of the bed, and from the assumed mean pitch of the rims. The closer these two figures of elevation were to each other, the safer was the projection. In addition to the preceding, other data obtained by the survey coincided in locating the channel line nearer to the north line of contact. Finally, a satisfactory projection having been made, the shortest line of a tunnel was located from the north face of the ridge, the entrance being in a sharp, precipitous ravine.

"The running of the tunnel showed that the true location of the channel line had been very closely approximated to at that point by the trial projection. The first upraise made at the 2400-foot station broke through the bedrock directly against the cement, about 40 feet up. The tunnel was then continued to the 3450-foot station, and another upraise made. This last disclosed so great a rise in the bedrock from the first upraise that a third upraise was made at the 2000-foot station, which at 15 feet up broke through the bedrock into gravel containing gold in paying quantities. This point was on the north edge of the channel. Further development located the center about the 2100-foot station. The success of this work led to the application of the same method of engineering investigation to a study of what is locally known in Placer County as the Forest Hill Divide.

"Though the surveys and necessary investigation are by no means complete yet, sufficient has already been established to prove that the channel just described as being found in the Mountain Gate and Red Point mines is continuous through almost the entire length of the Forest Hill Divide, from Hogs Back, ten miles northeast of Damascus, to Peckham Hill, three miles southeast of Todds Valley, a total distance by the line of the channel of about thirty miles. On this channel, in addition to the already noted mines, are undoubtedly the Turkey Hill Mine (now closed), the Paragon at Bath, the Dardanelles below Forest Hill, the celebrated Mayflower, two miles north of that town, and the Gray Eagle at Spring Garden, two miles west of Todds Valley. These at this time are the mines producing; in addition are many other claims in a more or less undeveloped condition, and some practically worked out after yielding enormous amounts of gold. Among the undeveloped or partially developed mines are the Hogs Back Consolidated, Indian Springs, Golden Fleece, Adams & Sellier, Georgia Consolidated, Baker Divide, Excelsior, Mountain, Spring Garden, and Big Channel. Among those that have yielded largely, but that are not worked to any extent, are the Gove, Maine, Independent, Rough and Ready, Jenny Lind, and Mountain, all lying under the town of Forest Hill. This main channel seems to have had many tributaries, all rich in gold, and all, so far as determined, coming in from the east. On them are located many mines that have yielded large amounts of gold, notably those above Michigan Bluff and on the Deadwood Divide. At the present time the Dam Mine, south of the Red Point, is the most important.

"The advantages of the engineering method of channel location over the uncertain haphazard work of the early miners, are such as warrant its application in every locality where drift mining is carried on, and further, to the examination of all unprospected ground in which it is possible for an auriferous gravel deposit to exist. By its use it is possible to determine in advance of doing any underground work on the claim:

1. The approximate location of the line of the bed of the channel.
2. The approximate elevation of the bed at any desired point.

3. The location of inlets, outlets, or breakouts, and a reasonable certainty of distinguishing between them.

4. If inlet, outlet, or breakout, the probable length of rim to be run through and the depth at which it must be penetrated.

5. If no inlet, outlet, or breakout, then the nearest point of the channel line to the surface for tunnel or shaft, as may be most desirable.

6. The determination of the size of the channel and the probable extent of its pay lead. This being of advantage in estimating the probability of yield sufficient to warrant the necessary outlay of capital in development.

7. The tunnel or shaft can be so located as to have the shortest possible length or depth with the greatest possible certainty of finding the channel, and of thereafter being permanently utilizable for working the mine.

8. The preceding make it possible to estimate in advance the probable expenditure that will be necessary to open up the mine, and to avoid any unnecessary expenditure, thus assuring the greatest possible economy of opening and of working.

"The entire drift mining district is covered with the evidences of failures to reach the buried gold-bearing gravels. Hundreds of tunnels, slopes, and shafts, abandoned after the expenditure of thousands of dollars, are silent witnesses to the inefficiency of the practice of the early miners in their search for the auriferous gravel deposits. In one example coming under the observation of the writer, a tunnel was run 800 feet into the bedrock of a mountain, the tunnel entrance being immediately over a small channel, when an engineer's examination would have shown, first, that the location of the channel was at the starting point; second, that in the line of the tunnel there was no channel for two miles; and third, that the cement relied on as a channel indication was only a shell ten feet or so in thickness lying on the sloping bedrock, the great mass of it having been eroded by the present river cañon. On the same channel, containing gold only in sufficient quantity to justify development on a small scale commensurate with its probable yield, through imperfect knowledge of it, was expended in the aggregate $150,000. After the expenditure of the larger portion of this amount, an engineering examination disclosed, what it could equally well have done before, that the expenditure of only a fraction of that amount was warranted by the probability of return; and with reference to the work done, that certainly $60,000 of expenditure could have been saved. The gold yield would very nearly, if not entirely, have balanced the other $90,000 expended.

"The cost of an engineering examination by the methods described varies, dependent on the circumstances of the particular mining claim it is desired to develop, from $500 to $5,000, sometimes exceeding the latter figure. Under ordinary conditions $1,000 is a safe estimate for the services of a competent engineer and the necessary field assistance. So interdependent are all of the drift mines, that the owners of adjacent mines, from the inspiration of self-interest, should not hesitate to render all assistance in their power to this kind of investigation. The cost of running a tunnel will average, under all conditions, about $12 a foot, so that the saving of 80 feet in the length of a tunnel will balance $1,000 of expenditure for an engineering investigation. The sinking of a deep shaft will average upward of $30 a foot—a saving of 33 feet in depth will balance the same $1,000. Aside from this, the certainty that the first tunnel will develop the ground, and be sufficiently low to drain it, will counterbalance several thousand dollars expense of preliminary investigation. In sinking a shaft, the knowledge of the probable depth carries with it the possibility of adjusting from the start the hoisting and pumping plant to the ultimate possible demands that may be made on it. In the development of a mine, in the knowledge of the writer, a large sum was expended in the sinking of prospect shafts, and a further much larger amount, unnecessarily, in a main working shaft, by reason of adding piecemeal to the hoisting and pumping plant from time to time, as the demands on it increased, all of which, in the aggregate, amounting to at least $10,000, might have been saved by the expenditure of $500 for a preliminary engineering investigation, for which the conditions were more than ordinarily favorable."

Having discovered the location and extent of the mine by prospecting, and the probable yield of its pay lead having been estimated, the question of costs is the next thing for discussion. Cost is made up of the following items:

1. Prospecting, which includes all the preliminary work.

2. The cost of opening the main working tunnel or shaft, its maintenance and extension, and the cost of all working appliances and buildings.

3. The cost of mining, which includes the cost of drifts and gangways, breasting out the gravel, conveying it to the surface, timbering the workings, ventilating and draining.

4. The cost of obtaining the gold from the gravel after it is brought to the surface.

The first two of these items may be called the construction account, which is closed when the mine is ready to produce. The last two items are the cost of mining or running expenses.

The construction account should aggregate as small an amount as possible, consistent with the proper opening of the mine. For the permanent working opening of the mine a tunnel, shaft, or slope is necessary. The tunnel is the most advisable construction, where possible. It should be run so as to be under the bed of the channel at the lowest point which it is designed to mine. Working up stream in a channel with a uniform grade, this main tunnel can usually be run on the surface of the bedrock. Working down stream, the tunnel is run in the bedrock underneath or in the rim to one side of the channel. In cases where an adit tunnel is necessary to reach the line of the channel, its direction should be at right angles to the presumed line until it reaches it, whence its direction and grade are controlled by the channel to a considerable extent. Grades may vary from one-fourth of an inch to a rod to ten inches to a rod, the practice being determined by the character of the track rail, the weight of the cars, and the power employed to take them in and out of the mine.

Dimensions of the tunnel vary greatly. This depends altogether upon the conditions which govern the mining operation. A main working tunnel may have dimensions of eight to ten feet in width at the bottom and seven to eight feet in clear height. The difference in the cost of timbering between a large tunnel and a small one is very little. The speed with which a tunnel can be run and its cost are dependent upon the kind of rock penetrated. Closer estimates can be made on the cost of tunnels run by machine drill than those run by hand.

The final charge against a construction account is for the surface plant to recover the gold. Should the gravel be soft and uncemented, a dump with sluices and a water supply, under a small head, constitute the plant. Should the gravel prove hard and cemented, it may be worked in the stamp batteries of a quartz mill. The gravel dump is usually piled near the tunnel entrance. It consists of a heavy planked floor, having two slopes, one from the sides inward to a sluice box running through the middle, the other with the grade of the sluice box. This flooring should be sheathed so that it can be renewed when worn out. From the edges of the flooring, walls are built up, the area of the flooring and the height of the walls being regulated by the desired capacity of the dump. For mines from which large amounts of gravel are taken daily, a discharging arrangement for the cars at the dump is of great advantage.

The sluice boxes beginning in the dump are continued beyond it a convenient distance and discharge into the ravine, in which are placed additional boxes, undercurrents, and tailing dams, all designed to recover the fine gold that passes the initial string of boxes. The special utility of the dams is in impounding the tailings until they are slacked sufficiently to free the gold cemented to the matrix. The boxes are 12 x 16, or 24 inches wide and of equal depth, and have a grade of 10 to 18 inches in twelve feet, depending upon the character of the values and the amount of water used. The bottoms are fitted with riffle bars that can be removed and reset rapidly.

The water supply is usually provided for by reservoiring the mine drainage, as well as what surface drainage is available. From this reservoir, an iron pipe goes to the lower end of the dump and is controlled with a gate. The nozzle for sluicing the dump is connected with this pipe by canvas hose for the purpose of flexibility.

The opening and working of a drift mine through a shaft is only advisable under conditions which make a tunnel impracticable. It may be that a shaft of no great depth, comparatively cheap and rapid of construction, is possible where it would require a very costly tunnel to reach the same gravel deposit.

"Where the mine must be opened by a shaft it is advisable to make the construction thorough and permanent from the beginning. The shaft point being located by the preliminary engineering investigation and prospecting, the sinking is done, as far as practicable, with horse power hoisting gear, the influx of water being taken out by the bucket as far down as it can be done without delaying the work of sinking. This point will be from 40 to 140 feet in depth from the surface. The power plant for hoisting and pumping is then set up, being proportioned, so far as hoisting is concerned, to lifting the gravel from the estimated depth the shaft will have when completed, and for pumping the probable amount of water that may be encountered when the mine is fully opened, a considerable margin of safety being advisable in providing for this, so that there will be no straining of the machinery. An additional margin of power is provided for to secure ventilation. Of course, wherever obtainable, water power is used, being far more economical. Most of the existing plants are, however, steam, the shafts being on the summits of the ridges, where it is not possible to get the necessary pressure for use of water.

"While sinking the shaft and prospecting, a bucket can be used to best advantage, in removing excavated material; afterwards in mining a cage on which a car can be lifted is preferable, as it saves one and possibly two handlings of the gravel. The preferable style of pump is the Cornish, both in sinking, as most rapidly adjustable to the conditions of changing depth, and afterwards as being able to control an increased flow by increase of speed alone, and as having less liability to breakages. The influx of water comes from the several gravel or diluvial strata passed through in sinking. This can be cut off from the bottom of the shaft by sumps, and pump stations placed where the flow is cut by the shaft. The shaft should be built in two compartments, one for the hoist and the other for the pump and man-way. It is timbered with framed square timbers, lagged on the outside and boarded on the inside in the hoisting compartment. The size in the clear is four and one-half by nine feet or five by ten feet. The size of the framed timbers is eight, ten, twelve, or fourteen inches square, and the sets are placed four, five, or six feet between centers, as controlled by the character of the ground passed through. The lagging is two inches thick. In lava there is no strain on the shaft, but some of the gravel and sand strata cut through are more or less liable to loosen and some of the slightly indurated clays are apt to swell. The cost of sinking a shaft can be safely estimated for the first fifty feet, $10 a foot; for the second fifty feet, $20 to $30 a foot; for the next one hundred feet, including the power, hoisting, and pumping plant, $50 to $60 a foot, and thence up to four hundred feet depth, from $60 to $75 a foot.

"Though a favorite method of opening a mine by the first drift miners, a slope is the least advisable now, and would only be employed under special conditions of economy, as, for example, in working from a flat, too extensive to be tunneled under, to a channel underneath a precipitous mountain slope, which would involve too deep a vertical shaft; or in mining from an inlet where neither tunnel nor shaft is practicable. Usually the conditions that indicate a slope as the most direct method of opening the mine can be better satisfied by a shaft and thence a tunnel from its bottom. In practice, if an extensive body of pay gravel is developed by a shaft or slope, a tunnel is subsequently run to mine it. This was done in the Derbec Mine, near North Bloomfield, Nevada County, a shaft 367 feet deep and a steam power plant being replaced by a 2000-foot tunnel. Also in the Mayflower Mine, at Forest Hill, a tunnel nearly 6000 feet long has been run to replace the shaft through which the discovery was made. At the present time no drift mine in the state is being worked through a shaft. A few are prospecting through shafts, with the intention of running tunnels if a sufficient amount of pay gravel is developed. The surface arrangements for working the gravel after it comes from the mine are the same in the case of either shaft or slope as already described for a tunnel.

"The preceding pages have considered the dead work of development specially chargeable to construction account. With its completion this account is closed, and all subsequent work and expenditure is a charge in a new account: the running expense of working the mine. The expediency of the expenditure of the capital used in the construction account must be determined on in advance from the results of the preliminary engineering investigation and prospecting. Once laid out, its return, as before noted, is from the net yield of the mine over its running expenses. These running expenses come under the several heads, as follows:

1. Opening up the channel or pay lead by main tunnel, drifts, and gangways. Prospecting for pay lead when it is lost.
2. Breaking out the pay gravel.
3. Timbering.
4. Drainage.
5. Ventilation.
6. Track, switches, upraises, and dumps in the mine.
7. Cars and motive power for moving the gravel out of the mine.
8. Working the gravel after being taken out of the mine.

"The main tunnel, when in the channel and pay lead, is constructed in larger dimensions and more carefully than the drifts and gangways only intended for temporary service. If timbered, the best timbers are used and the work of setting them up is done so as not to require early removal. In hard, cemented gravel, requiring blasting, the drilling is single or double-handed, power drills not being used. In wide channels, as a precaution against possible caving, a pillar of solid ground is left on each side of the tunnel, from twenty to forty feet wide, dependent on the stability of the ground. Where the working tunnel is in the bedrock underneath, following the line of the channel, the pillar need not be left, as the tunnel in the gravel becomes a main drift for only temporary use in mining the ground between its connections with the bedrock tunnel. These connections are made every 200 to 400 feet, as determined by convenience of working. The main tunnel in the gravel on the bedrock, and also the bedrock tunnel, are sometimes affected by the swelling of the bedrock, usually upward in direction. Under such circumstances very heavy timbering is advisable, and the floor of the tunnel must be cut down from time to time in order to keep it from closing up. The necessary excavation can be done without interrupting the working of the mine or the use of the tunnel, as the swelling rock is always soft and can be worked out without using powder. The main tunnel is kept as straight as possible and in the center or lowest depression of the channel. Drifts are run from it at right angles to the rims of the channel or the limits of the pay lead. These are timbered and lagged in soft ground, but in not as permanent a manner as the main tunnel. The distance apart of these drifts is not governed by any special rule. Both the main tunnel and cross drifts are used to prospect the ground and locate the pay gravel. This will control the distance to some extent, but not absolutely. In the pay lead the distance apart is decided so as to secure the greatest convenience of working.

"In the Red Point and Hidden Treasure mines, in which the pay leads are very wide, the drifts are 120 feet apart. In wide channels these drifts are connected by gangways parallel to the main tunnel, the practice in their number and distance apart being equally flexible. In the Red Point Mine they are run 65 feet apart, thus blocking out the ground to be mined into rectangles 120 feet by 65 feet. In the Bald Mountain Mine, at Forest City, the practice was to run both the drifts and gangway 80 feet apart, leaving a pillar of 40 feet to protect the main tunnel. In the Hidden Treasure Mine only one gangway is run, connecting the ends of the drifts at the extreme limits of the pay lead as determined by the prospecting of the gravel from the drifts. This difference in practice is accountable for by the difference between the character of the mining ground in the several mines. In the Red Point it is hard and compact, and the openings, except in the breasts, require no timbering; the gravel, however, is not regular in the amount of the gold it contains, and closer prospecting is advisable to cut out ground too poor to pay for mining. In the Hidden Treasure the gravel is soft, the bedrock swells, and every opening requires timbering to protect it; therefore only absolutely essential openings are made for working, the gravel being so uniform in gold yield that special close prospecting is not needed. The Bald Mountain gravel was soft and as regular in yield as in the Hidden Treasure. The smaller blocks were doubtless made to facilitate the convenience of working, only four and one-half feet of depth of ground being taken out. The cost of a main tunnel in the gravel drifts and gangways naturally has a considerable range as between different mines, but is practically constant in the same mine. In hard, compact gravel, requiring blasting, the cost of main tunnel, six by seven feet, will be from $4 to $7 a foot; of drifts and gangways five by six feet or six by six feet, from $3 to $5 a foot. In gravel not as difficult to drill as the preceding, but still requiring blasting, $3 to $4 for main tunnel, and $1.75 to $3 for drifts and gangways. In soft gravel, requiring timbering, the figures are about the same as those last given, the greater penetrability being offset by the expense of timbering. In some mines, particularly where pay gravel on the high rock of the rims is being prospected for, the drift for this purpose is run as wide as sixteen feet and as low as four feet in height, in order to cover as much ground as possible, and move as little waste. This method is, however, unsystematic, and not to be recommended for large mining operations.

"In connection with the opening of the mine and mining properly belongs the consideration of the utility of trained engineering skill to drift mining. Already in this connection, with the preliminary work of location of the development works, has been shown the value of this skill. In connection with the permanent opening and subsequent working it is of equal service and value. Every drift mine should have an accurate working map, on a scale of twenty or forty feet to an inch, of its underground workings and their connections with the surface. On this map should be shown the tunnels, shafts, gangways, rims of the channel, and blocks of ground cut out for breaking down; also the location of air and water pipes and connections. On it can also be placed the figures of the estimated yield of the different blocks of unbroken ground, as determined from the prospecting and the figures of actual yield after working. The ground worked out from week to week can be marked on the plat by shading. A map so made is of great service in directing the main tunnel and prospecting drifts in advance of the ground being mined out, and in making air and working connections. To facilitate the surveying, the underground foreman should set points at all angles and intersections in tunnels, drifts, and gangways. These are best set overhead in the roof, as less likely to be disturbed by the mining operators. A wooden plug is first firmly driven in a short drill or gad hole, into this a ring or hook from which a plumb or lamp can be suspended.

"The breaking out of the gravel, the mine being opened as described, is done from the faces of the gangways, or if there be none, from the faces of the drifts or main tunnel. In the Hidden Treasure Mine the side of the gangway toward the main tunnel is the working breast, and is broken down by the miners working the whole length of it at once to a distance of eight feet from the gangway. A new set of posts, parallel to the gangway, with caps and top lagging, is put in, timbering up the ground to the face. The track is then moved from the gangway close up to the breast, and more ground broken out as described. Not all of the gravel is taken out of the mine, but only the fine, the boulders being piled back on the ground from which the track has been moved. A block is thus worked up to the line of the pillar thirty feet from the tunnel. The gravel being soft, no powder is used. From one to two feet of soft bedrock and three to four feet of the gravel are mined out. The method described was also used in the Bald Mountain Mine, and, in fact, is employed in all the large, systematically operated mines, where the gravel is broken out without powder, and where the bedrock is soft and of comparatively even surface. In mines where the gravel is hard, and has to be broken out with powder, there is no special care taken to keep the breast faced up even, it being most economically broken down by working from the corners of the blocks. The drifts and gangways are kept open and the track is not changed, but the broken out gravel is shoveled out from the breast to the cars after separating the large boulders. To make the shoveling easier temporary plank floors are put in. In ground with hard bedrock of very irregular surface the track is not moved up to the working face, but the gravel is shoveled out to it, as already described. In narrow channels, where the cross drifts are not connected by gangways, the sides of the drifts and main tunnel, preferably the former, are made the working breasts, and the ground broken out from them.

"A method in common use among many of the miners for working their ground, even where the pay lead is quite wide, is to mine its entire width in one semicircular or curved breast without running a main tunnel ahead or cross drifts from it. The unsystematic and unnecessarily costly character of this method is evident. There is no opportunity to protect the ground in advance of working it, and there are no reserves to keep up the output while exploring through a barren portion of the lead. Timbering can not be done systematically; the main tunnel in which the track runs is not permanently protected against caves. If the ground is hard there is great waste of labor in drilling and of powder in breaking. The method is to be condemned under any conditions.

"In blasting gravel in a drift mine the object is not only to get the full effect of the powder in the amount of gravel thrown down, but to pulverize it as completely as possible so as to free, in a measure, the gold. The best practice is to drill deep holes three to five feet, chamber them, and then put all of the powder in the bottom of the hole, tamping it in tight. Experience will determine the proper load for the holes to produce the best results. A slow burning powder is preferable, proportioned to the tightness of the ground. The usual strength used is from 30 per cent to 40 per cent nitroglycerine.

"The output of gravel from a drift mine is measured by carloads; the size of the car is not, however, uniform, so a comparison must take this difference into account. The cost of breaking out the gravel independent of the expenses of handling it afterward, or those connected with the opening of the drifts and gangways, timbering, and track, is controlled by the hardness of the gravel, expense for powder and candles, and the rate of miners' wages. The tabulated figures will show the cost in several mines, and furnish fair comparative data for estimating:

Name of mine	Candles, per ton.	Powder, per ton.	Carloads, per pick.	Weight, car load, tons.	Total weight broken out.	Rate miners' wages.	Cost per carload.	Cost per ton.
Dardanelles	$0 02	$0 40	1.50	1.30	1.95	$3 00	$2 55	$1 95
Paragon	01¼	--------	1.35	2.00	2.70	3 00	2 25	1 72½
Red Point	01¼	25	2.70	1.00	2.70	3 00	1 27½	1 27½
Hidden Treasure	01	--------	4.30	1.00	4.30	3 00	71	71
Manzanita	01¾	--------	4.30	1.40	6.82	2 50	52	57

"The Dardanelles gravel is hard, but it is worked in the curved form of breast condemned above. The Paragon gravel is soft, but worked out in an irregular manner only slightly improved over the preceding. The Red Point gravel is as hard as that in the Dardanelles, but the systematic method of mining employed makes it cost considerably less per ton. The Hidden Treasure and Manzanita gravels are soft, but completely unlike, the first named having a white quartz gravel, and the last a fine quartz gravel, with a large amount of sand and no waste to speak of. If the surface of bedrock is hard and left unbroken on breasting out the gravel, it is cleaned thoroughly, the crevices and surface being scraped with a special tool to remove every particle of gold, before the boulder waste is thrown back on it. One, or at the most, two men can clean the bedrock of a large mine as fast as it can be uncovered by the breast miners.

"The timbers used in a drift mine are obtained from the surface of the claim or from timber land near at hand. They are usually cut by contract and delivered at so much apiece—8, 10, 12 and 15 cents, dependent on the size, and the distance they have to be hauled for delivery. The posts, caps, and sills are cut from cedar, sugar or yellow pine, or spruce, and are relatively valuable in the order given. They are rough cut in length from 6 to 8 feet, or more or less if so required, and from 8 inches to 14 inches square, rough hewn or split, or, particularly in the larger sizes, left in their natural shape with only the bark removed. The lagging is split from the same varieties of wood, one and a half to two and a half by six inches by six feet. It is cut by contract at so much per thousand—$8 to $12. Both timbers and lagging are allowed to dry and season where cut out of the trees. Before being taken into the mine they are dressed and cut by the carpenter for the special purpose for which they are to be used. The carpenter also prepares large numbers of wedges, which are used to brace the timbers into position.

"The different timbermen in the mines each has his peculiar method of framing the tunnel sets of timbers, and all seem equally efficient. In placing a set in position, seats are cut in the floor for the posts to rest in, sills being rarely used. The cap is mortised at the ends into which the top of the posts fit. These sets are placed from three to six feet between centers, dependent on the solidity of the ground they support. The lagging is driven in behind the timbers and towards the face. If not perfectly solid in place, wedges are driven in, always pointing towards the face. The size of the tunnel timbers and the inclination of the posts from the vertical depend on the ground, the size of the tunnel, and desired permanence of the work. The Bald Mountain tunnel was in soft ground; round timbers twelve to sixteen inches in diameter were used; the sets placed four feet between centers, and the inward inclination of the posts, two feet nine inches in a rise of six feet six inches, the tunnel being nine feet, three and one-half feet, and six and one-half feet clear dimensions. The usual sized timber employed is eight to ten inches in diameter, and the inclination of posts, one foot three inches to six feet six inches. In the cross drifts lighter posts are used, but the sets are framed and the construction similar to the main tunnel, except that only the top is lagged, unless the ground should be very soft and sliding. In the working gangways the posts are set vertical and the mortised end of the cap only covers half the top of the post, so that as subsequently other parts are set up in breaking out the ground as described, the same post will support the end of a second cap. Only the top is lagged. In breaking out the ground the timbering is similar in kind and construction to that in the gangways. In hard ground that is broken out by blasting comparatively little or no timbering is necessary. As a rule only posts are used, with or without caps, the latter not necessarily supported by more than one post. The caps are secured in place by wedges driven towards the working face to prevent the roof starting, and as a protection to the post during blasting. The actual cost of setting up timbers and lagging in working ground of a mine can not readily be segregated from the cost of breaking out the gravel, as it is not usually done by a special force, but by the miners themselves under the direction of the underground foreman.

"Drainage may or may not be a most important item of expense in working a drift mine. It is practically nothing, where the mine is worked on an ascending grade through a tunnel. If it has to be lifted, some of the various pumping devices are used. For a shaft, the Cornish pump has already been referred to and commended. For making short lifts, less than twenty feet in the altitudes at which most of the drift mines are located, a siphon can sometimes be employed, and requires but little attention. Underground, direct acting steam pumps are usually employed, the steam being brought from the surface, necessarily with a loss of power dependent on the distance. In the Mayflower Mine, at Forest Hill, Placer County, while worked through the shaft, six pumps were employed underground, some of them 2000 feet from the boiler supplying the steam. In addition to the great expense involved, it made a most unsafe plant, for the breaking down of one of the pumps destroyed the efficiency of all beyond it in lower levels of the mine. Sometimes water power is available for direct use underground.

"In the Mountain Gate Mine, at Damascus, Placer County, the main bedrock tunnel is 40 feet lower than the channel being mined, at the point where it is cut off by the deeper blue lead channel. The channel descending inward to this point, all its drainage, about forty inches, was collected at the inner end and utilized to run a forty-foot overshot water wheel, which gave power to pump and hoist from the cross channel sixty feet lower.

"In the Turkey Hill Mine, near Michigan Bluff, water was brought in through the air shaft, 300 feet deep, falling on an overshot water wheel, and furnishing power to pump the water from the deep working to the level of the main tunnel, whence it reached the tunnel entrance by natural flow. In this instance only a fraction of the effective power of the water was utilized, as it was allowed to fall free in the shaft. Another device that can be used efficiently if water power under high pressure can be had, is an arrangement on the principle of the hydraulic elevator and Bunsen pump, by which the water power is used directly to obtain a suction and elevating force. The cost of putting in the plant for the water power devices is not very large, and they can be operated cheaply, the last described at only the cost of the water for power, as it requires, once started, no attention whatever. The water power pumps require more or less repairing and examination to keep working well. The first cost of steam pumps and their connections and the expense of operating are a heavy charge on the gravel mined. Their employment is condemned except for very rich ground that can be drained by no other means."

Ventilation of drift mines is important. With only one opening, the ventilation is by a fan at the tunnel entrance forcing fresh air to the working face by a pipe line or withdrawing the vitiated air by the same means. If water power is available the blower can be driven by it at little expense. Air shafts and drifts are constructed at minimum cost for ventilation.

The main tunnel and drifts all have track laid in them and switches are placed at all junctions and intersections. In the main tunnel the track is permanently laid. In the drifts and gangways it is removed as soon as the ground is worked out and relaid elsewhere. Light rail is sometimes used and sometimes strap-iron spiked on wooden stringers,

Cars vary greatly in dimensions, shape, material, discharge gear, and capacity. In small mines the cars are moved by hand; in larger ones by a horse or locomotive.

"The treatment of the gravel to obtain the gold is either by washing it from the dump through the sluices, or, should it be cemented, crushing it in the stamp batteries of a quartz mill. The washing plant has already been described. In small mines where not over one hundred carloads a day are taken out, the washing is done by the superintendent or the foreman. In larger mines there are one or two men steadily employed at the washing and cleaning up of the sluices. The latter is done in sections, the upper boxes certainly once or twice a day in some mines—those further away from the dump less frequently. The tailings are not allowed to escape at once, but are caught in brush and log dams, and allowed to accumulate and slack for several months, when they are rewashed. The common practice in the mines is to sell them outright for a lump sum to Chinese or others, who take the chances on getting back their cost and the expense of washing. The cost of washing per carload is from 1½ cents (with large amounts of gravel and free water) to 3 cents. In milling gravel the batteries are best fed by automatic machine feeders. Hand labor is necessary, however, to separate large cobbles, which can be partially screened out by a grizzly. The cost of milling gravel per ton in the Paragon Mine, with steam power mill, is $0.35 a ton. At the Dardanelles, the cost with steam power, five-stamp mill, is $0.33 a ton. With water power mill the cost of milling the same gravel (exclusive of the cost of the water) would be $0.20 a ton.

"The specialized duties in which labor is employed in a drift mine, the ratios between the amount of labor in these several capacities, and its cost, are seldom exactly the same in any two mines. The figures for three mines are given here, and will furnish a fair basis from which to make estimates for projected work:

	Dardanelles		Paragon		Red Point	
	Number	Wages, per day	Number	Wages, per day	Number	Wages, per day
Foreman	1	$3 50	1	$3 50	1	
Shift bosses	2	3 00			2	$3 00
Breasters, white	24	3 00	17	3 00	9	3 00
Breasters, Chinese					9	1 75
Tunnel men			2	3 00	2	
Drift men, white					3	3 00
Carmen, inside, white						
Carmen, inside, Chinese					6	1 50
Carmen, outside	1	3 00	1	2 50		
Drivers					2	3 00
Blacksmith	1	3 00	1	3 00	1	3 50
Blacksmith helpers					1	3 00
Carpenters					3	4 00
Surface men, white	1	2 50			1	3 00
Surface men, Chinese					4	1 75
Engineers	2	3 50	1	3 50	2	3 50
Battery feeders	2	2 50	2	3 00		
Totals	34		25		46	

"In the Dardanelles the full force possible is working in two shifts. The force can be doubled in the Paragon, the present force only working days. At the Red Point, with sufficient water for washing, a hundred additional men could be employed in breaking out ground already opened by drifts and gangways. The Hidden Treasure Mine employs from 100 to 175 men in all capacities, the larger proportion in breasting out gravel. In the Bald Mountain Mine as high as 250 men have been employed at one time in all capacities. In the running expense of a drift mine, the cost of labor is by far the largest item, the proportion, as compared with all other expenses, being nearly uniform in all of the mines. For the Hidden Treasure Mine the ratio for four months of 1888, taken at random from the books, is: For wages 78 per cent; all other expenses, 22 per cent. For the same mine for the eleven years from 1877 to 1887, inclusive, the ratio was 78 per cent, and other expenses 22 per cent.

"The considerations that should govern the developments of a drift mine and the most successful practical working of it, summarized from the preceding pages, are the following: The development or opening of the mine should be done in the manner that will make the subsequent mining of the ground—that is, the running expense per unit (carload or ton) of gravel—the cheapest. This means, drift mining ground not being uniform in gold yield, that the greatest amount and area of ground can then be mined at a profit over running expenses, and that more thorough prospecting that is not dead work can be done. This points, not to the lowest possible construction account necessarily, but to that which will in the running expense of the opened mine make all the several items take their minimum value, and permit of the largest proportion of the total of all of them being expended in the actual mining or breaking out of the pay gravel. It is from this last consideration that the tunnel is the best form of opening, for through it can be reduced to their minimum value the several items of drainage, ventilation, and moving of gravel to the surface.

"The gold yield of the gravel is estimated at so much per carload, but the differences in capacity of the cars used in different mines makes direct comparison impossible. For convenience it is desirable to adopt as a unit the ton of 2000 pounds. The minimum limit of yield which it will pay to mine, or rather the minimum of the running expense of mining (for it will pay to mine gravel which will just meet this expense, as increasing the probability of discovery of richer ground), has a wide range as between the different mines. Probably the lowest paying gravel and the cheapest mining is that of the Hidden Treasure Mine. From February 27, 1888, to June 30, 1888—108 working days—the figures are:

	Per load, one ton	Total
Gold yield	$1.2347	$39,821 53
Wages	.7202	23,528 00
Contracts	.1077	3,464 78
Expense, material, etc.	.0957	3,086 94
Total expense	$0.9236	$30,079 72
Profit	$0.3111	$9,741 81

Number of days' labor, 11,164.50. Number of carloads gravel, 32,252.
For the eleven years, 1877 to 1887, inclusive:

Receipts.

Gold yield	$879,523 27
Receipts from other sources	19,176 16
Total	$898,699 43

Expenditures.

Wages	$490,297 64
All other expenses	137,064 35
Dividends	268,092 00

"The cost per carload ($0.9236) is exceptionally low, as under ordinarily favorable conditions $1.50 to $1.75 a carload is as low a figure as can be anticipated, and in most of the mines the cost is from $2 to $3."

In the above abstract from Mr. Dunn's report, it has been necessary on account of space limitations to omit much material of value, principally data referring to costs at several of the drift mines which were operating thirty-five years ago, when the article was written. Costs have since changed materially, but the relative proportion and weight of labor and material costs is about the same. For this reason the same factor of increase may be applied in estimating probable costs at the present day.

Methods of drift mining have not changed materially except in the application of modern mining machinery to tunnel driving. In the recovery of gold from the gravels, perhaps the only important change has been the adaptation of mills of the Price, and Krogh type to the moderately cemented gravels. Stamp milling is no longer in general use, unless the gravel is exceedingly tightly cemented. Mills of this type (the Price, and Krogh type) consist of a revolving hexagonal or octagonal barrel of about five feet inside diameter, and length varying from six to ten feet. The walls of this revolving trommel or screen are perforated, usually with slots of sufficient mesh to allow the fine material to escape, the boulders and pebbles in the gravel itself acting as a crushing medium until they are discharged. This mill is usually mounted on trunnions in such a way that if driven by water power through a belt, the water used on the Pelton wheel is discharged into the mill to serve as washwater.

In gravel which is not too tightly cemented, this type of mill gives excellent results and a very complete separation of the gold from the gravel without the necessity of crushing the entire mass and consequently at a much lower cost than the old-fashioned stamp mill.

MECHANICAL HANDLING.

In many localities irregular distribution and depth of the gravels, the inaccessibilty of water, character and type of bedrock, or the remoteness from transportation make necessary a departure from standard mining practices, and the adaptation of mechanical means for handling gravel in smaller units than those hitherto discussed.

There are innumerable methods which have been devised and are still being devised, but in general it may be stated that gravel which is handled by small mechanical units must of necessity be much higher grade than that which is capable of being handled in large volume by standard dredge or hydraulic methods.

The first method discussed will be that of handling gravel by elevator or mechanical conveyor. Attempts have been made from time to time to overcome the lack of sufficient dump in certain gravel deposits by means of mechanical elevators of the stacker or belt type. These machines have been expensive in their first cost, erection, maintenance, and operation. They have never been, so far as known, a financial success. However, inasmuch as no serious mechanical problems are involved, it seems reasonable that a successful machine of this type might possibly be designed, constructed, and operated. It should be lighter than most of the attempts made to date, and consequently more easily moved.

The most elaborate example of this type that has been constructed in California is the plant of the Tarr Mining Company at Smartsville. The gravel was hydraulicked to a sump where it was lifted by a bucket ladder, containing fifty-two 7-foot buckets, to a trommel where it was screened. The undersize went to the riffles and the oversize to a belt conveyor 570 feet long. At the end of the conveyor two Bleichert tramways were constructed to provide a larger dump for the tailings. This plant proved a failure, due largely to the cemented character of the gravel and the complicated nature of the machine, which required very much power to run.

At Beauceville, Canada, a mechanical elevator was erected by the New York Engineering Company at a cost of $25,000. Its lifting height was 50 feet from the bottom of the sump to the top of the bucket-line. It was equipped with 150 h.p., of which the bucket-line of $3\frac{1}{2}$ cubic ft. close-connected buckets consumed 50 h.p.; the remaining 100 being used to operate a 14-inch centrifugal pump. Its capacity was 2000 cubic yards daily. Cost per yard was about 12 cents. The failure of this plant was due to the high cost of operation and the low grade of the gravel.

Steam shovels of various types have been used in the mechanical handling of gravels. Where there is insufficient water for hydraulicking and where the gravel is more or less cemented, attempts to work ground by this method have been frequent but without any great measure of success.

As a digging machine, the steam shovel is very efficient, but its failure in gravel mining has been largely due to its lack of mobility and to the fact that its action is intermittent; that is, it dumps its loads suddenly, and the sluices are in consequence alternately over-burdened and over-flooded. This method of charging sluices is conducive to loss of fine gold. Difficulty has always been experienced in keeping the sluices up to the working face of the machine. In the old type of machine which laid its own track the shovel had all the disadvantages of the dredge in that its heavier parts were very cumbersome. Recently, however, a type of shovel has been devised which runs on a caterpillar track, and it seems that this type is far better adapted to placer mining than any shovel hitherto devised, because of its greater flexibility.

In connection with this development of the steam shovel, the power-driven belt-conveyor has recently been used. On gravel in compara-

PHOTO No. 15.   Yukon Gold Company's Mechanical Elevator, Dawson,
Yukon Territory.

tively accessible districts, which runs 20 cents per yard or better, there is a possibility of using this type of machine which is well worth looking into. At the present time a company is operating low-dump gravels in the bed of the Calaveras River by this method. The caterpillar steam shovel, with its radius of action, due to the quickness with which it can be moved, will cut and deliver gravel of this type at great speed, and with a five or six foot conveyor-belt to carry the material up to a hopper at the head of elevated sluice boxes, it is possible to handle this type of gravel much more cheaply than formerly.

Many types of drag-scrapers have been used for the handling of gold-bearing gravel. In Alaska on Twin Creek the writer has seen frozen ground handled by this means at a cost not to exceed 50 cents a yard. A central steam power plant, operated by burning wood, was set in such a way that the drag scoop, traveling the full length of the cut, would dump into a car which was hoisted to a hopper at the head of the sluices. The sluices had been raised to give ample dump.

So many improvements have recently been made in drag-line scrapers that they are now very efficient digging machines in loose ground. A drag-line scraper at Cape Girardeau, in Missouri, several years ago dug a canal 200 ft. wide and 20 ft. deep in this type of material at a cost of 5 cents per cubic yard. It was equipped with a 6-cubic-yard bucket and had a capacity of 4000 cubic yards in 24 hours.

Other types of excavators used in handling gravel, such as the Hadsell, which operates from a central tower, and another unnamed type, which operates with variable-speed motors, have been successful. In this latter type a slow-speed motor is used for digging at the moment that the scraper takes its charge, and a faster motor is used to convey it to the tower where it dumps its burden into the sluices.

Photo No. 16. Drag Scraper near Fairbanks, Alaska.

Simple derricks with clam-shell or orange-peel buckets have often been used for this type of work, but the gravel has to be very high-grade to justify the expense of this sort of operation.

Various mechanical contraptions with the name of dry-land dredgers have been built and experimented with at different times. These usually consist of some form of bucket elevator mounted on wheels, and in many parts of our State still stand as monuments to the practical incapacity of their inventors.

The use of the centrifugal pump in handling gravel has already been commented upon, but there are certain areas of gravel still available along the Sierra foothills in regions which were once the deltas of the Cretaceous and Tertiary river systems, which are capable of being handled by this method. Where the foothills are low and rolling and reservoirs can readily be installed, it is possible to increase the head in the operations of giants supplied by these reservoirs by stepping a centrifugal pump into the line, provided the power costs are not too expensive. Probably gravel would have to run at least 25 cents a yard to allow any profit at all by this means of working. The use of centrifugal pumps as elevators for gravel has already been discussed under the head of dredging.

GROUND SLUICING METHODS.

The original and simplest form of mining was by means of the pan and the rocker. From the development of the rocker or cradle came the longtom, which was a sort of end-shake sluice which permitted the handling of more material than the rocker. After this came the development of the fixed sluice.

There exist deposits where whole areas of rolling ground are covered with a thin film of varying depths, which at times feathers out to bedrock, and upon which water can be brought but not under pressure. When the intervening gulches have sufficient grade for sluicing and disposal of the tailings, ground sluicing is the proper method. This method of mining has perhaps produced more of the world's gold than any other. During the early days in California, when gold production was at its maximum, by far the larger proportion of the gold was won by this method. In viewing the ancient Spanish workings in Colombia, in Peru, and in Bolivia, one can not fail to be impressed by the enormous amount of material that has been mined by this means.

The method consists of bringing the ditch to a point on the edge of the bank to be worked and allowing the water to cascade over the working face. The work of the water is assisted by the use of picks and shovels, and the larger boulders close to bedrock are worked around, if they can not be moved. The duty of water in this method is naturally low. For example, in a mine of this type in Colombia, 135 miner's inches, running over a 10-foot bank gave a duty of only 50 cubic yards per day.

What is known as booming or flooding is an alteration of ground sluicing methods, which is adaptable under certain conditions. Where the gravel is readily cut and does not attain a depth of over 50 feet and has a preferable average depth of 10 to 15 feet, this method is practicable provided that the slope of the bedrock is at least 7 inches to the rod.

A reservoir is built above the deposit in which a self-dumping, or self-opening gate is placed. This gate is so arranged, generally with an overflow box, that when the water rises to a certain height in the reservoir the gate opens and empties the reservoir with a rush, after which it automatically closes. The water pouring down the gulch carries away immense quantities of material until it loses its velocity, and by this means a large amount of the top gravel is readily stripped, leaving the bottom part, with the bedrock to be worked off by ordinary ground sluicing methods.

Where the gold is fine and generally deposited through the gravel, this method is not advisable, but where the gold is coarse and concentrated in a shallow streak or near bedrock, this is often an economical method of working.

DRY WASHING.

There are many parts of California in which the gold-bearing gravels are inaccessible to water in any quantity. As these gravels often run as high as $1 a yard and in some cases better than this, all sorts of methods of handling them without the use of water have been tried.

To the writer's knowledge, no absolutely dry process has ever been successful in handling this type of gravel, with the exception of a Mexican hand bellows or dry-washing machine.

The deposits are usually shallow, more or less cemented, and their gold contents are very unevenly distributed. Probably the best-known absolutely dry process is what is known as the Stebbins-Quinner machine. In most cases the gravel is first pulverized and screened, and the fine material is run over the Stebbins tables, which use air as a medium of concentration instead of water. The writer has yet to learn that any machine using air as a means of concentration will prove successful in handling gravel. Very often laboratory tests will show encouraging results, but it is found in practice that the so-called dry material is never absolutely dry but contains enough moisture to agglomerate on the canvas, thus interfering with proper concentration.

The hand machine of the Mexican, which has been used extensively in southern California and in Arizona, is an exception to the general rule regarding air separation machines, for the reason that it is portable, and the extremely spotted nature of most of the dry placers results in local concentrations, which for small widths and in quantities of two or three yards at a time will often pay fair day wages. Again the writer wishes to state with particular force that air separation processes, when applied on any large scale, have invariably resulted in total and absolute failure. Two conspicuous monuments to this failure still stand in the California desert; one at Goler, and the other at Coolgardie. Limited capacity and imperfect separation are the main reasons. Other machines of this character are so cumbrous that they lose the chief advantage of the hand machine, which is its ability to be transported from one spot or concentration of gold to another as soon as the pay has been worked out.

Where a limited amount of water is available, however, it is quite possible and feasible to work these dry placers, provided the ground is rich enough. The most successful method of handling this material that has come to the writer's attention is one which was tried out in the Randsburg district some time ago. It consists in the use of a very little water and the constant saving and return of the same for re-washing. In this method a sluice about 7 ft. long and 14 in. wide is given a head motion by an eccentric and is driven by a two-horsepower engine. The sluice has a false bottom with holes immediately behind the riffles, which are of the Hungarian type. The material is wheeled and shoveled in after the boulders have been screened out. The water carries out the fines on about a 12-inch grade into a wheelbarrow which stands over a pit. The wheelbarrow is punched full of holes and the water is drained into the pit from it. It is then pumped back by a centrifugal pump, which is run by the same gas engine from a countershaft through the false bottom, where it jets up behind the riffles to perform its service in carrying the material through the sluice again. The only loss of water is through evaporation and absorption. The riffles are always kept open by the uplifting jets of water, and the concentrating motion of the eccentric head makes a very effective jig.

Another machine which is used in localities where there is very little water has been developed by Mr. J. B. Giffen, of Sacramento. The machine consists of either a shaking or stationary hopper and water feed under pressure in the hopper, a shaking table with a side motion, set at a pitch averaging about 3 inches to the foot, and a type of riffle known as the Giffen riffle. The inventor claims that near Manhattan, Nevada, he could wash 40 yards of dry gravel with a supply of 9 gallons of water per minute, for 6 hours a day, the consumption being 80 gallons per yard. By the use of a trommel, the amount of gravel washed is materially increased. The dirt is fed into the hopper by an elevator. As the dirt dumps and spreads out, it is met by the water discharge, which thus gives a fair washing before the gravel reaches

PHOTO No. 17. View of Giffen Placer Machine. Taken when washing 24 yards per hour, at Rocklin, Cal.

the table. The upper 18 inches of the table has another riffle, thus allowing the dirt to spread and cross the table in a thin wide stream, which causes a very fair degree of concentration to occur before the first riffle is reached. Four sections of riffle each 16 inches long were used, although it was only necessary to clean the upper one daily, the second every other day and the other two once a week. Where water is scarce, it is pumped back and used over and over. The plant can be moved and put in operation in less than one-half a day. The inventor claims that under ordinary circumstances the ground can be delivered, washed, and the tailings cleared away, so far as necessary, for 30 to 35 cents per yard.

BLACK SANDS.

The subject of black sand mining has been very ably and thoroughly covered by Mr. C. A. Logan in Bulletin No. 85 of the State Mining Bureau. For this reason only a very brief résumé of the methods of recovery in common use will be given in this chapter.

One of the commonest machines in use is the Kellogg black sand machine. This device consists of an inverted funnel with pockets around the circumference. As the sand and water are poured upon this, a boiling action occurs in the pockets and a sort of concentration is brought about. This is a very simple machine and appears to have been of considerable use in handling the black sands of the Oroville district.

Perhaps the simplest type of machine used on the beach sands consists of a plain duck board riffle, fastened to a plank bottom, which is set out near the edge of the sea as the tide is coming in, and is kept constantly in such a position that the outgoing waves would wash the gold-bearing black sands over it. This machine has been used near Crescent City, and has paid day wages to the men operating it. Possibly one of the best types of concentrators for this sort of work is the Huelsdonk, which the writer has seen operating in Sierra County, and

PHOTO No. 18. Giffen Placer Machine. General appearance of plant.

which has successfully been used for years in recovering platinum, gold, amalgam, and mercury from black sand concentrates at the La Grange dredge. The description given by Logan[1] of this plant follows:

"The concentrator works under still water in a box or trough which is 16 feet long, one foot wide inside, and about one foot deep, being made from two-inch planks. A small gas engine mounted on the sluice furnishes power for shaking the screen and the concentrator, and for pumping water. The shaking motion is given by an eccentric with $\frac{3}{8}$-inch travel. The screen moves on a single bolt support on each side, and the power is applied against springs. From the screen the sand and water pass on to an apron which extends one-half the length of the sluice and is perforated at regular intervals so as to distribute the sand along the table proper. This apron and the table are bolted together and are shaken at the rate of 180 r. p. m. They travel on rollers along the bottom of the sluice, and require little power. The table proper is essentially a long narrow galvanized-iron covered trough, extending the full length of box, and tapering at the lower end to a groove scarcely $\frac{1}{4}$-inch wide and deep. The sand enters the groove at the upper end and as the shaking motion forces it along the lighter constituents are crowded to the top and forced over the side, falling into a bottom compartment which shakes with the table and which can be used to give a middling, or to discharge tailing. The concentrate travels the length of the groove and is tapped off through a spigot at the end. Middling and tailing are tapped from the side near the end. Huelsdonk

[1]Logan, C. A., Platinum and allied metals in California; Cal. State Min. Bur., Bull. 85, pp. 100-101, 1919.

claims the unit can handle two cubic yards of gravel or one ton of mill tailings an hour. Twenty cubic yards of gravel give two gallons of concentrate. The concentration with mill tailings is said to be 100 to 1. At the La Grange dredge 8 tons of black sand concentrate were reduced to about one-third of a gold-pan full, which contained the year's output of platinum. Only ¾-horsepower is said to be required for the concentrator. The demonstrating model has a 1½-horsepower engine, which is claimed to be more than ample for pumping water and operation. A one-inch centrifugal pump gives an ample supply of water.

"The saving by this machine appears to be very satisfactory, and the concentrator seems to have a wide field of application, but ought to appeal especially to the small miner or the man who wants a portable outfit which is easy to operate and requires little water. The installation complete, including engine and pump, weighs 600 pounds, and the heaviest part is the engine. Two men are required to run the outfit where hand shoveling is done."

Another machine, of the airblast type, has been developed by Sutton, Steele and Steele, of Dallas, Texas. This machine has been used with considerable success on concentration of light and finely distributed materials. Whether it is adaptable to heavy black sands with the variety of heavy concentrates which are produced from this class of material, seems still to be undetermined, but from the looks of the blueprint submitted to the writer, it looks as if it had possibilities.

METHODS OF PROSPECTING.

In the discussion of the different methods of operation of gold placers, various factors have been mentioned which determine the choice of the method of work. In the case of the proper examination of alluvial properties with the idea of determining the method of working to be adopted, the examining engineer will carefully consider all of these factors and their relative importance one to another. If the enterprise is to be hydraulic, the question of water supply is paramount, and it may be necessary to determine the extent of the available water, the amount of the annual rainfall, and its flow-off, absorption and evaporation; also all data pertaining to reservoir sites and ditch lines. As a survey of this kind runs into considerable money, it is often well to delay its undertaking until at least some of the ground has been sampled. In the case of a dredging proposition, the determination of the amount of available water-power may be held over in the same way.

In addition to the factors already mentioned, the following points should be investigated: First, all questions of title, royalty, duty on mining machinery, and of the existence of legal obstacles to regular operations.

The determination of the valuable content of the property in question, that is sampling, upon which hinges the crux of the examination, calls for the most painstaking and systematic care on the part of the engineer. He must determine as exactly as possible the tenor of the ground and the yardage available for working.

Gold in gravel exists in two forms: that which for want of a better name, we call the 'gravel-gold,' which was deposited simultaneously with the gravel, and that known as 'flood-gold.'

Economically the former is all important, as it forms the chief supply of gold in alluvial deposits. It is usually laid down once for all, and covered by subsequent layers of material. Generally it occurs in well defined bands, or channels, of varying width and thickness. When it occurs coarse, the richest portion is usually found on or near bedrock. Strata of sand are generally barren. Surface soil may contain enough gold to warrant the cost of its removal.

Flood-gold is the bane of the engineer. The majority of cases brought to his notice by the uninitiated, for example, deposits where natives are making a living wage by surface panning, prove on examination to be nothing but barren gravel beneath a thin film of flood-gold. Economically this form of gold is of small moment. It is usually a finely divided flour or flakes derived either from the attrition of the lighter part of the gravel-gold, or being deposited at the present day by the living stream from quartz veins within the drainage area. It is of a vagrant nature, and migrates with each seasonal flood. It is found concentrated on heads of islands or bars, and entangled in the vegetation covering the flood-plane. It is purely superficial, and never attains any concentration in depth.

The gold in the rivers of Europe is usually of this character; notably in Hungary, where the peasants, after cleaning up the river carefully, can depend upon a new 'crop' of gold in the succeeding year. Here the rivers, whether worked or not, show no augmentation of the gold from year to year by reason of these 'crops.' The rivers of Spain are also notorious for this character of gold, as also are many rivers in tropical countries. Deposits capped with flood-gold may or may not contain gravel-gold beneath; therefore its presence is no criterion of concentration in depth.

High banks of gravel, suitable for hydraulicking, can sometimes be sampled by channelling their surface after the manner in which veins are sampled. Usually the bank is divided into panels of one or two hundred feet wide. After cleaning the face, a cut is commenced as near bedrock as possible, and extended up to about 5 feet. This cut is conveniently about a foot wide and about 6 inches deep. Exact measurements should be taken of the length and breadth and depth of the cut in several places, and the cubic contents of the sample in place carefully determined. Material from the cut should be measured loose, and carefully washed in a rocker; the gold is then amalgamated, parted, and weighed, and the value per cubic yard, based on the volume in place, is computed.

Commencing at the level of the top of the first cut, another sample should be begun and the operation repeated as far up as possible. It is not necessary that all the divisions of any one sample be in the same vertical line. They may be 'staggered' to suit convenience. By this method the exposed surface of a cliff of gravel can be systematically sampled; greater facility being afforded if old tailing is heaped at the base. This method gives accurate results. During an examination of this character in Peru occasion was had to check certain cuts by resampling, and it was found that the yield checked almost exactly. To determine the gold content of the gravel at any distance back from the face, and also the pitch of the bedrock, either shaft-sinking or drilling must be employed.

Formerly, in California, it was the practice to sample a deposit by first installing a small hydraulic plant for the purpose of washing a portion of it. If the clean-up was satisfactory, the property was approved. This method has, happily, fallen into disuse, because it gives misleading results. It gives a large sample from one place, which may or may not be representative of the whole deposit. A case is known where $1,100 taken out by the experimental giant on a run of a few hundred cubic yards was subsequently only equalled by a whole season's working of 200,000 cubic yards.

Small dredges have sometimes been employed to test dredging-ground, especially in New Zealand. Where funds are available, this method gives most satisfactory results, but is bound to be expensive in the first cost. Subsequent expense may be offset by the clean-ups.

It is difficult, if not impossible, to determine the amount of the sample taken in gravel that is not firm enough to stand when channelled. In sampling gravel of this character, the sample should be weighed and its volume in place calculated. Before commencing to sample, either by pits or channelling, a deposit of gravel of this character, several trial-tests should be made in order to determine the weight of a cubic yard of gravel in place. The amount of water in these tests should, as nearly as possible, correspond with that in the subsequent samples.

Of all the methods of testing alluvial ground, that by means of pitting, or shaft-sinking, is the most reliable. Any alluvial ground is not properly and finally tested without the sinking of at least a few shafts. As will be pointed out later, drilling, in the writer's estimation, is only a makeshift, the value of which, aside from preliminary testing, is dependent upon checking by shafts.

In order to save the handling of excessive material, and to keep down the cost, the shaft should be as small as possible. In California, where the ground does not have to be timbered, a size of 2 ft. 3 in. by 4 ft. is often employed. This gives one cubic yard of material to every three feet of depth. Where convenient, all of the material is put through a rocker. Sometimes channels are cut down the sides and the gravel from the measured cuts is washed. It is hardly necessary to state that gravel taken from cuts should not be quartered down, or 'coned,' as in vein-sampling, for the gold will be concentrated by each re-handling, and erroneous results obtained.

Herewith is the working drawing of a rocker, used with excellent results in prospecting ground. Most rockers, the designs of which have been published, are too high and too short, and consequently are not good savers of fine gold.

The cost of shaft-sinking, compared with that of drilling, is high. Nevertheless, this high cost is justifiable in consideration of increased confidence in the accuracy of the results obtained. A shaft not only determines the value of the gravel, but the character of the ground can be determined in much greater detail than in case of the drill. Boulders lying in loose gravel, often missed by the drill, are exposed. Bedrock can also be cleaned and its character ascertained.

Almost all of the dredging grounds in California have been tested both before the erection of dredges, and afterward, as a guide to operation, by the Keystone No. 3 traction-drill. This drill uses a 6-inch casing and is operated by steam. It weighs about 6 tons, and costs, with all tools complete, a trifle under $2,000. It operates a string of tools on a cable attached to a drum, the tools being raised by the pressure of a walking-beam, and dropped by gravity. The drill and stem weigh about 800 lb. The crushed core is removed from the casing by means of a sand-pump of the vacuum type, raised and lowered by an independent reel and cable. The present-day practice in drilling with this machine is to drive the casing one or two feet by means of a driving-block clamped on the drill-stem; then by means of the drill to crush the material forced into the pipe, and pump it out, leaving a core depending on the character of the gravel. This core should be of such thickness as to prevent an inrush of material from below; and yet it should not be thick enough to form a plug that will force away gravel from the casing when driving is continued. The material from the pump is collected in a box, where it should be measured, and carefully panned or rocked afterward. The amount of core that has been pumped from the pipe is recorded in the log, and also the number and size of the 'colours,' or specks of gold, as well as the character of the ground.

In California gold is usually classified in three different sizes: the finest (No. 3), weight about 1 milligramme. Those running from 1 to 5 mg. are classed as No. 2; over 5 mg. as No. 1. The fine colours are usually aggregated and grouped as No. 3. This method of recording the approximate value of the various pumpings enables the engineer definitely to locate his 'pay-streaks.' In Alaska, a different classification is used, namely, No. 1, 2, 3, and nuggets; No. 3 is 1 mg.; No. 2 is 4 mg. and No. 1 is 8 mg., while colours above 10 mg. are called 'nuggets.'

After the hole has reached bedrock (into which it should penetrate), the casing is pulled by means of what is called a 'pulling-cap.' A cap screwed on the top of the upper casing is hammered upward by means of a heavy steel bar, which passes through the cap within the pipe, and is operated by the walking-beam and cable. The operation of pulling pipe in the Keystone is sometimes serious, consuming a good deal of time and occasionally resulting in loss of casing, due to the stripping of the threads.

In drilling with the Keystone, it is not considered good practice to drill below the casing, for particles of gold lodging beneath loose flat rocks, even outside the area of the bore, are shaken down into the pocket formed by the drill and subsequently taken up by the sand-pump. On the other hand, it is not good practice to have too thick a core in the pipe, as there is then a tendency to force material to the side of the path of the descending casing-material, which rightly belongs in the core. With the Keystone, this usually only happens in light running ground. As a general rule the tendency of the Keystone is toward an excess of core.

Employing the outside diameter of the casing as a basis for calculating the volume, has been found in practice to give high results. A pipe formula reducing the value of these results, and based on check shaft-tests made by W. H. Radford, has been generally accepted. This formula reduces the results obtained by the theoretical formula about $14\frac{1}{2}\%$. In its simplest form, it is as follows:

$$\text{Value of gravel per cu. yd.} = \frac{\text{Value of gold obtained} \times 100}{\text{Depth of hole in feet.}}$$

It has been stated already that the Keystone is apt to extract an excessive core. In order to correct this excess, the value of the gold is reduced proportionally by the ratio of the number of inches driven in the 'pay-streak' to the number of inches in the excessive core. When the excess is in barren material, no correction need be made.

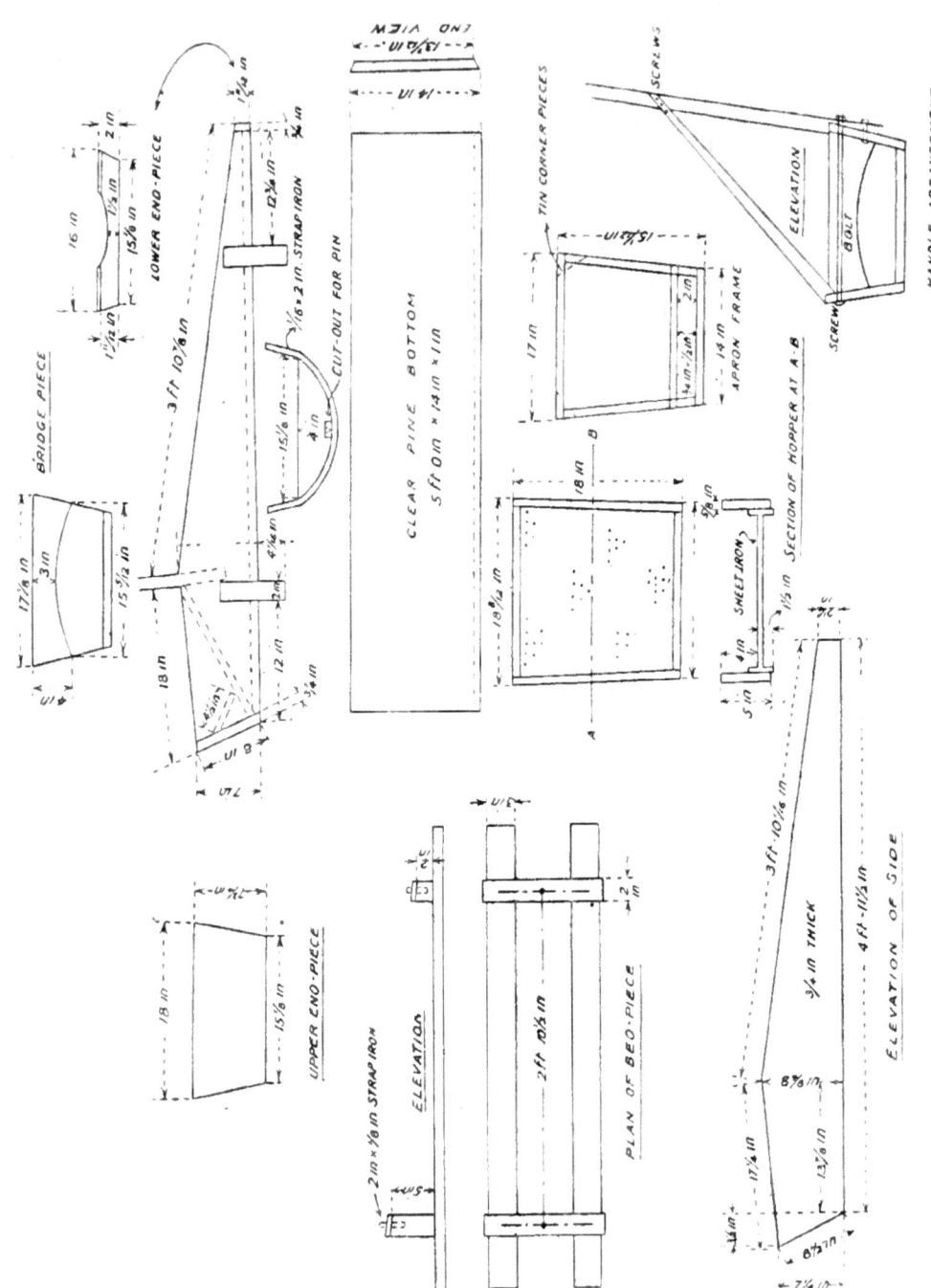

PLATE I. Details of Rocker.

Owing to the fact that the casing remains stationary during pumping, excess core can be accurately measured in the Keystone. In this respect it differs from the Banka hand-drill, as will be shown later.

At Oroville, California, the Keystone drill has been employed for so many years, and so much of the ground tested by it has been subsequently dredged, that it is possible to determine within narrow limits the ratio between actual dredge-recovery and averages indicated by drilling. Where drilling has been carefully done, 75 to 80% of the estimated yield can be recovered by the dredge. This includes all losses; not only those in the tailing, but unrecoverable islands and corners left behind in the course of operations. In Alaska, however, where the ground is shallow, the gold coarse, and on a shattered bedrock, dredging results have usually exceeded the estimated yield, the percentage of recovery being from 103 to 198. When these high recoveries have been obtained, the dredging depth has always been greater than the drilled depth, in cases by as much as 30 to 40 per cent.

Alluvial gold is found occurring under such widely different conditions, conditions that affect both drilling and dredging alike, that no arbitrary factor can be applied in estimating the yield of a deposit from the results obtained by drilling. In general, the choice of any discount factor must be guided by experience. Special conditions govern each case, such as the fineness of the gold, the proportion of clay, the number of boulders, the method by which the deposit is to be worked, and, finally, the personal equation of the drilling-crew.

The need of a portable drill in testing alluvial ground has led to the adoption of several types of drills operated by hand, some of which are patterned after the well-digger's tools. Mr. Newton B. Knox had occasion to test a dredging property, in Korea, that consisted of fine gravel about 20 ft. deep, resting on a soft clay bedrock. The gold was concentrated in a narrow streak from 6 to 12 inches thick, resting upon the bedrock. There being no drills available, one had to be devised on the spot. Attempts were made to drill by following the Keystone method, that is, to drive the pipe, break the core, and pump, but this had to be abandoned owing to the ineffectiveness of the home-made pump. The following method was finally adopted:

Holes were bored in the overburden by means of an auger to a depth of five or six feet, and a 3½-in. pipe (the only size available) was inserted. This pipe was shod with a sleeve. A sleeve sawed in half, and screwed on top, served as a driving cap. The pipe was driven into the ground with a wooden driving-block operated by coolies standing upon a movable platform supported by ladders. While being driven, the pipe was rotated by means of wooden clamps or chain tongs. No attempt was made to pump out the core. The pipe was driven into the bedrock to a depth of about two feet, after which it was pulled by means of a wooden tripod and chain-blocks, or block and tackle, the pipe being rotated during pulling.

The material was then removed from the pipe by means of a long corkscrew auger. Occasionally the pipe picked up a boulder in its descent, when the hole had to be abandoned and another started. Otherwise the drill worked satisfactorily. About 175 holes were put down by this means, which gave a fair value to the ground. The owners of the property afterward had it re-drilled with a machine of the Banka type, and the result of 200 holes tallied almost exactly with the home-made drill.

One of the best-known types of hand-drills that is at present in prac-tical use is the Banka type, which was invented by a Dutch engineer in the year 1858 for the purpose of testing tin deposits in the Dutch East Indies. This drill is at present manufactured under its original name by a firm in Haarlem, and is also manufactured in the United States by a New York firm, with some modifications of the type, under the name of the Empire Drill.

The form usually employed is that of a 4-in. flush-jointed casing shod with a steel-toothed cutting-shoe. The platform, upon which four men stand while operating the string of tools, is attached to the top of the casing. The various tools used in drilling or pumping are raised and lowered by means of a steel rod 1 in. square, which is jointed in lengths corresponding with the lengths of casing, which are usually five feet. The theory of this machine is that the casing will sink under the combined weight of the men and tools while the platform and casing are being rotated by either man or horse power. The cutting action of the revolving shoe, while sinking, is claimed to cut a core which is representative of the gravel passed through. The results of the writer's experience with this drill have not been entirely in accordance with the claims put forward by its American makers and, in fact, difficulties have been encountered that appear to have been entirely overlooked in the discussion of the drill in the maker's pamphlets.

One of the chief points of variance is the method of sinking. It is claimed that the drill will sink largely by rotation alone. In attempt-ing to sink by this method, and keeping a careful record of the cores obtained results have been as follows:

First: In ground containing any quantity of gravel about the size of a man's fist or over, no progress whatever is made by pure rotation, and eventually driving has to be employed.

Second: In ground consisting mainly of fine gravel, no progress is made until the pump has reached the bottom of the casing. Then, while pumping with little or no core, sinking proceeds with fair rapidity, but in measuring the core obtained it is found to be excessive. For this reason it was observed that good progress can only be made by rotation when the material is pumped away from under the shoe, a process which necessarily gives excessive cores. The best progress that can be made in sinking, consistent with the obtaining of an accurate core, is made by driving with a battering ram, approximately a foot at a time. The ram should not be too heavy as better core is obtained with a light ram.

Another claim made for this drill is that "when running ground is encountered with the rotated pipe there is much less danger of getting an incorrect sample, because the rotated pipe immediately sinks through the gravel till it strikes solid material." In the writer's experience the presence of running ground has always made itself felt by an influx into the pipe; and while sinking to more solid material an excess core has always been picked up.

One of the weak features of this drill is the method of removing the core from the pipe, that is, with a ball-valve pump. In breaking gravel to any size within the pipe, either by means of a pump or the ordinary breaking-bit, a certain amount of the core is always driven out of the casing and lost. This results in a deficient core. In shallow gravel this tendency is more marked than in holes over 20 ft. deep.

Another cause of deficient core is the peculiar shape assumed by a worn shoe. In contrast with the Keystone shoe, which is bevelled toward the outside and tends to retain its shape, as the casing is not rotated, the shoe of the hand-drill, which is at the start perfectly flush with the casing inside and out, soon wears to a tapered shape with the greatest bevel on the outside. This, of course, results in forcing aside material that properly belongs to the core.

The actual value of all drilling results is entirely dependent upon the representative nature of the core obtained. If an excess of material from the productive layer is secured, unless allowance is made for that excess, the results obtained from the hole will be proportionately high. A deficient core from the productive layer, on the other hand, gives results unfair to the ground.

The ordinary method of keeping a core record is to compare the total measured volume of core obtained, upon completion of the hole, with the theoretical volume required by a cross-sectional area of the casing and the depth driven. This practice gives a false sense of accuracy, because a deficient core at one stage of the drilling is often balanced by an excessive core at another. If the excess happens to come from the rich layer, the results will be too high, although the total measured core checks the theoretical. A driller, knowing that his core was deficient at one point could counteract this by deliberately taking an excess at another, and yet the records would fail to show the incorrect nature of his drilling.

In order to overcome this difficulty and to keep a record of the cores obtained at each pumping, all the material derived from each successive foot of sinking should be measured carefully in a measuring box and compared with the theoretical volume, based on the actual sinking of the casing. Thus at any time it is possible to tell whether excessive or deficient cores are being obtained. The extra work involved in this is negligible and does not interfere with the work of the panner.

There is undoubtedly a slight expansion of volume of the gravel after it has been drilled over its natural volume in place. The amount of expansion of gravel under these conditions, that is, after it has been broken up, re-deposited under water, and drained, has been determined in various ways. To some of the statements made with regard to this expansion, the writer took exception while prospecting some ground in Colombia. One of these statements was that gravel that has been excavated occupies from 30 to 40 per cent more volume than when in place. This might possibly be true in the case of loose gravel, but for gravel that is broken up, washed, and laid down again under water, it is certainly entirely too high. The result of the test made by the writer in conjunction with Mr. Knox was as follows:

Volume of gravel in place	Volume of gravel after washing measured cu. ft.	Ratio of expansion %	Kind of gravel
0.48 x 0.78 x 1.37 ft.=0.513 cu. ft._____	0.518	101	Very fine gravel and sand
0.5 x 0.92 x 1.6 ft.=0.736 cu. ft._____	0.848	115	Medium gravel
0.925 x 0.474 x 1.43 ft.=0.627 cu. ft.___	0.690	110	Medium gravel

Considering the above tests to be reasonably accurate, it appears that an ideal working of the drill without excessive or deficient core should give in ordinary gravel 110 to 115 per cent of the theoretical core. By careful drilling, the writer has found that from 90 to 100 per cent of the theoretical value can usually be obtained.

As a result then of employing the theoretical pipe formula and using the gold recovered from the above amount of material, it is seen that the calculated results will have been automatically discounted from 10 to 25 per cent. This is proved by the fact that the above tests show an actual expansion of from 10 to 15 per cent, whereas careful drilling gives but 90 to 100 per cent of the theoretical volume. This discount does for the hand drill what the Radford or Keystone formula does for the power drill, as the latter formula discounts by nearly 15 per cent.

The formula usually employed in calculating results with the Banka drill for gold consists in dividing the value of the gold recovered, multiplied by 240, by the depth of the hole in feet to obtain the value of the gravel per cubic yard.

The Banka drills have come into vogue for testing gold placers so recently that the results obtained upon ground that has been subsequently dredged are not very extensively proved. When figuring the actual dredgable value of the ground, it seems advisable to employ at least the same discount of 20 to 25 per cent on results obtained with the hand drill, as is usual in the case of the Keystone.

In comparing hand and power drills it seems logical to state that each type of drill has its own particular scope. The Keystone is to be preferred in places where the ground is heavy, where roads are fairly good, where labor is expensive, although skilled, and where repair shops are available. These conditions exist in most parts of California, and the power drill is employed extensively there. On the other hand, where transportation is a problem, where labor is cheap, and the ground is fairly light, the hand drill seems to be the logical instrument. The main advantage of the Keystone over the Banka type is that it can drill stiffer ground and deeper holes. It requires for operation a panner, a driller, and a fireman, as well as a man and team to haul water, fuel and casing. Its disadvantages are its higher cost, its weight, its liability to expensive repairs and its cumbrousness in moving over heavy, wet, or rugged ground. In pulling pipe, the threads are often stripped, resulting in loss of casing, and at times the operation of pulling is a long and tedious one. The former difficulty could readily be obviated by the adoption of flush joints and heavier threads.

The Banka type of hand drill is less expensive; it is light, weighing about a ton, and is portable. It requires few, if any, repairs, and with the exception of the renewal of shoes, not many extra parts. The drill requires a panner, four men on the platform, and a man and horse for motive power. If a horse is not available, men must be employed for the work of turning. In deep holes it usually requires at least eight men to turn the drill. In Colombia, in using this type of drill, the writer found that a crew of 12 men afforded an opportunity for the men on the platform to change frequently with the men on the rotating sweeps. It would have been impossible to work one crew of four men continuously on the platform for ten hours per day, but by working them in alternating shifts so that they were only working on the platform for one hour out of three, much better speed was obtained.

As an illustration of hand-drill limitations, Mr. Newton B. Knox speaks of testing some ground in southern Siberia. The deposit consisted of loose gravel from 9 to 15 ft. in depth, containing some large stones and resting on a hard upturned shattered sandstone bedrock. The gold was exceptionally coarse, nuggets as large as peas being common, and it occurred generally deep within the crevices, joints, and bedding-planes of the bedrock, quite out of reach of the drill. Some of the gold was found beneath boulders and bedded in the ground and lying upon bedrock. As would be expected, the deposit was patchy. Fifteen holes, systematically laid out, were drilled and they gave discouraging and erratic results. To test as far as possible the reliability of the drills a series of holes was put down across a valley midway between shafts that had been sunk during the winter. These shafts were fifty feet apart and had given results of 20¢ to $1.20 per cubic yard. Though the holes were sunk in the bedrock, the drill failed to find any gold whatever, and consequently this method of testing was discontinued.

Another type of drill in more or less common use on the Pacific coast and in Alaska, which obviates some of the difficulties of the heavier Keystone power drill is that known as the Union. This drill is driven by a light gasoline engine and is mounted in such a way that it can be readily transported on a wheeled frame. This drill has been extensively used in Alaska and has given very satisfactory results where the ground is not too deep. It will undoubtedly drill ground which is much heavier and stiffer than can be drilled satisfactorily by the Banka type.

While in actual operation the Keystone drill sinks at a faster rate than the hand drill and yet, if the time lost in moving is considered, the hand drill averages about the same number of feet per day. In ordinary gravel two 60-ft. holes a week is a good average for the hand drill. The pulling of the casing in one of these holes ordinarily requires only about an hour.

In Keystone drilling in California and Oregon, in ordinarily stiff gravel, about 30 ft. per day is a good average, but the time lost in pulling and moving reduces the weekly average to about the same as that of the hand drill.

In general the Keystone, the Union, and the Empire or Banka types are standard drills which are excellent in their respective fields, but each one requires for careful work considerable experience in operation. In order that confidence may be placed in any work done by drilling, it should be done conscientiously and cores obtained should be normal and representative, and the pipe formulas used should be applied with discretion. In the writer's opinion, at best all drilling is more or less of a make-shift and of far more use in determining bedrock conditions than in determining the values in the ground. For the latter purpose shaft sinking is far more accurate when the quantity of material taken out from the shaft is washed through a sluice box and the results obtained from this are later checked by channel cuts made on all four sides of the shaft and sampled independently.

The number and distribution of holes and shafts depend chiefly on the character of the deposit and the distribution of the gold. Of course the more holes per acre, the more accurate will be the results, but also the greater will be the cost of the examination. In the Oroville district, where the gold was more or less uniformly distributed through the gravel, as soon as the character of the deposit was known, one hole per acre was considered sufficient. In new and untried districts or where the ground shows a tendency toward spottiness, the holes should be closer together.

Where the gold occurs in channels, holes are placed at regular intervals from 200 to 500 ft. apart, with the intermediates put down wherever necessary to define the limits of the channel. The rows are usually started 1000 ft. apart, but this distance is determined by local conditions, and should be left to the discretion of the engineer in charge.

It is often possible to segregate and reject low-grade or barren areas, thereby reducing the yardage, but bringing up the average value for the remaining ground. In this way sufficient yardage may be proved to justify an installation, which would be impracticable if the whole was taken into consideration. Having proved the yardage and determined the character of the gravel, the type and size of installation can be decided. From operations under similar conditions, the working cost per yard can be fairly accurately estimated, and consequently the expected yearly profits can be determined within reasonable limits. Knowing the total life of the property, the profits to be derived, and the cost of equipment and installation, its present worth can be calculated. Comparing this with the cost of the property, its value as an investment may be reasonably approximated.

In all prospecting work, the factors of most importance are the experience, judgment, and discretion of the examing engineer. A wide range of experience will often be of more value in the interpretation of results obtained from an expensive drilling or shaft sinking campaign than any other factor in the examination, and the use of judgment in interpreting these results will often mean the difference between the absolute and unrecoverable loss of a large investment, or the profitable employment of the same in insuring a long and useful life for the property under consideration.

Bibliography.

Hydraulic Mining. Report U. S. Mng. Commission. Mining & Scientific Press, beginning Nov. 28, 1874, and running continuously through May 1, 1875.

Hydraulic Elevators. Mining & Scientific Press, July 21, 1877.

Hydraulic Mining. Mining & Scientific Press, beginning Oct. 13, 1877, and running continuously through Dec. 22, 1877.

Dry Placer Amalgamator. Mining & Scientific Press, Nov. 10, 1877.

First Hydraulic Elevator. Mining & Scientific Press, Nov. 6, 1880.

Drift Mining. Mining & Scientific Press, Jan. 7, 1882; Jan. 14, Jan. 21, Jan. 28 and Feb. 4, 1882.

Hydraulic Elevators. Mining & Scientific Press, Mar. 11, 1882; April 4, 1896; Dec. 1, 1894; Sept. 26, 1896.

Dredging. Mining & Scientific Press, Nov. 30, 1895; Sept. 4, 1897; Nov. 13, 1897; July 8, 1899; Jan. 27, 1900; Feb. 3, 1900; Dec. 15, 1900.

Hydraulic Mining. Mining & Scientific Press, Apr. 10, 1897 to June 12, 1897. Sept. 18, 1897; Dec. 18, 1897; Dec. 25, 1897. January 1, 1898 to Feb. 5, 1898. March 25, 1899; Mar. 10, 1900; June 14, 1902.

Placer Mining by Machinery. Mining & Scientific Press, Feb. 5, 1898.

Wing Dams. Mining & Scientific Press, Mar. 5, 1898.

Sampling Placers. Mining & Scientific Press June 17 1899; June 24, 1899; June 27, 1903; July 11, 1903; Feb. 3, 1906; Oct. 20, 1906.

Working Cement Gravels. Mining & Scientific Press, Oct. 27, 1900.
Placer Mining Methods. Mining & Scientific Press, Mar. 29, 1902; April 5, 1902;
  April 12, 1902.
Hydraulic Mining. Mining & Scientific Press, Nov. 16, 1901; Dec. 6, 1902; Aug. 5,
  1905; Aug. 12, 1905; Feb. 14, 1903; Apr. 18, 1903.
Drift Mining in Cement Gravels. Mining & Scientific Press, Jan. 3, 1903; Jan. 10,
  1903.
Ditches. Mining & Scientific Press, July 18, 1903.
Centrifugal Pumps in Hydraulicking. Mining & Scientific Press, April 2, 1902.
Flume Construction. Mining & Scientific Press, Oct. 22, 1904.
Drag Line Work. Mining & Scientific Press, Dec. 17, 1904.
Hydraulicking with Pumps. Mining & Scientific Press, May 12, 1906.
Hydraulicking Bedrock Cuts. Mining & Scientific Press, Dec. 1, 1906.
Loss of Gold in Placer Mining. Mining & Scientific Press, Feb. 23, 1907.
Pump Sluicing for Gold. Mining & Scientific Press, Feb. 13, 1909. July 3, 1909;
  Nov. 11, 1911.
Ditches. Mining & Scientific Press, Mar. 6, 1909; April 8, 1911.
Prospecting Placers. Mining & Scientific Press, May 22, 1909; Jan. 7, 1911; Jan.
  28, 1911; Feb. 25, 1911; July 6, 1912; May 9, 1914; Oct. 10,
  1914; July 4, 1914.
Nature of Gold in Alluvials. Mining & Scientific Press, May 29, 1909.
Art of Piping. Mining & Scientific Press, Nov. 13, 1909.
Hydraulicking Pipe Clay Gravels. Mining & Scientific Press, Jan. 22, 1910.
Stacker for Hydraulicking. Mining & Scientific Press, Feb. 19, 1910.
Hydraulicking in California. Mining & Scientfie Press, May 21, 1910; Dec. 2, 1911.
Dry Placer Machinery. Mining & Scientific Press, Nov. 12, 1910; July 13, 1912.
Losses in Hydraulic Mining. Mining & Scientific Press, Jan. 21, 1911.
Elevating with Ruble Grizzly. Mining & Scientific Press, April 13, 1912.
Pay Streak in Placers. Mining & Scientific Press, June 1, 1912.
Stacking Hydraulic Tailings. Mining & Scientific Press, Aug. 24, 1912.
Thawing Frozen Ground. Mining & Scientific Press Jan. 17, 1914.
Undercurrents. Mining & Scientific Press, June 5, 1915; May 1, 1915.
Prospecting Wet Placers. Mining & Scientific Press, Jan. 9, 1915.
Gold Saving on Dredgers. Mining & Scientific Press, Aug. 5, 1916.
Self Shooter. Mining & Scientific Press, Mar 17, 1917
Timbering in Deep Placers. Mining & Scientific Press, Aug. 11, 1917.
Dams. Trans. Am. Soc. Civil Engrs., Volume 57.
Log Dam Construction. Eng. & Min. Journal, Nov. 28, 1896; June 17, 1899.
Cement Gravel Mill Eng. & Min. Journal, May 1, 1897.
Notes on Hydraulic Mining. Eng. & Min. Journal, Oct. 30, 1897.
Sweeny Placer Machine. Eng. & Min. Journal, Mar. 26, 1898.
Evans Hydraulic Elevator. Eng. & Min. Journal, May 14, 1898.
Sampling Placers by Shaft. Eng. & Min. Journal, July 29, 1899.
Mining Lowgrade Placers. Eng. & Min. Journal, Nov. 11, 1899; Nov. 30, 1901.
Pump for Hydraulicking. Eng. & Min. Journal, March 7, 1903.
Hydraulic Mining. Eng. & Min. Journal, May 19, 1906; June 2, 1906; June 9,
  1906; Nov. 10, 1906; Nov. 17, 1906; Dec. 26, 1908; Apr. 8,
  1911; Mar. 13, 1915.
Prospecting. Eng. & Min. Journal, Dec. 21, 1907; Dec. 12, 1908; May 1, 1909;
  Mar. 12, 1910; Apr. 5, 1913; July 24, 1915; Aug. 21, 1915.
Ruble Elevator. Eng. & Min. Journal, Nov. 7, 1908; Dec. 18, 1909.
Saving Fine Gold. Eng. & Min. Journal, Aug. 26, 1911; Apr. 25, 1914.
Dry Washing. Eng. & Min. Journal, Oct. 17, 1914; Jan. 23, 1915.
Scraper for Gravel Mining. Eng. & Min. Journal, Aug. 14, 1915.
Gold Recovery in Placers. Eng. & Min. Journal, Sept. 18, 1915.
Hydraulic Mining. Second Annual Report, Cal. State Mining Bureau. Ninth
  Annual Report, Cal State Mining Bureau.
Drift Mining. Eighth Annual Report, Cal. State Mining Bureau.
Hydraulic Dredging. Trans. Am. Inst. Min. Engrs., Volume 40. Also Vol. 6.
Saving Fine Gold. Trans. Am. Inst. Min. Engrs., Volume 18.
Gold Dredging in California. Cal. State Min. Bur., Bulletins No. 36 and No. 57.

CHAPTER III.

PLACER RESOURCES.

The principal gold placers of California are located in the eastern half of the State in the Sierra Nevada Mountains between Lassen County on the north and Mariposa County on the south. It is from this area that the greater bulk of California gold has been produced and will be produced in the future. For this reason a map showing the location of the auriferous channels of this region has been prepared in considerable detail.

On the other hand, it must not be assumed that all California's gold has been produced from this area. A small proportion of it came from the dry placers in the south, and some of it from the upper section of the Kern River, as well as from various outlying districts in the Sierras and in the Coast Range. Second only in importance to the Sierra region is that portion of the State embraced in Del Norte, Humboldt, Siskiyou, and Trinity Counties, drained for the most part by tributaries of the Klamath River. The Smith River in Del Norte County also contributed a portion of the total.

Starting, then, in the northen portion of the State, a discussion of this region is the first that will be taken up. Although the Smith River is not a tributary of the Klamath, it will also be included in the same general heading.

SECTION I.

KLAMATH RIVER REGION.

Certain geological concepts should be clearly outlined before taking up the discussion of this region in detail, in order not only to consider the origin of the gold in this region, but also to distinguish between two distinct types of auriferous deposits which predominate; namely, the Cretaceous shore-line or conglomerate deposits; and the fluviatile deposits of the same period, together with all secondary concentrations of the same.

During the Cretaceous period, when the shore-line of the ocean extended east of the chain of islands, which is now the Coast Range, and its waters beat against the foothills of what is now the Sierra Nevada Range, there existed an island of almost continental dimensions close off the shore, which embraced in area territory which is now included in Humboldt, Siskiyou, Trinity, Shasta, Del Norte, and Tehama Counties in California, as well as a large portion of southwestern Oregon.

The shore-line of this island on the west, to a great degree, coincided with the present western coast line. Starting at Gold Bluff, its course can be traced northerly nearly to Requa, where it swings northeasterly and crosses the upper portion of Smith River; thence going northerly to about the line of the Rogue River in Oregon; thence easterly toward Grants Pass, and southerly down toward Cottonwood Creek and Henley in California; thence still further southerly down the Shasta Valley to Yreka, where a turn southeasterly carries it across Shasta County to Redding and almost down to Red Bluff; thence northwesterly again through Salt Creek and out to Big Lagoon and up to Gold Bluff again.

The drainage of this island was largely expressed in two rivers: one flowing northerly, and the other southerly. These gravels have been classed as those of the "second period of erosion." The southward flowing stream can be traced from the upper Trinity, through Minersville and the Weaverville Basin, out by way of La Grange to Hayfork and Hyampom. The northward flowing stream can be traced from the Salmon River in Siskiyou County, through Crapo Meadows and Portuguese Creek, on the Klamath, to the head of the Illinois River in Oregon, and thence out by way of Briggs Creek and Galice Creek to the northern shore-line. In addition to this, there were coastal streams of minor importance, the largest of which corresponded in location to the present drainage system of Clear Creek, whose ancient delta is expressed at Igo and Ono.

The shore-line gravels can be clearly traced through Siskiyou and Shasta counties, and have already received considerable discussion, notably in Mr. Russell Dunn's article on the "Auriferous Conglomerate in California," in the Twelfth Annual Report of the State Mineralogist of California. For many years this shore-line was regarded as a true river channel, whose behavior occasioned great confusion; but when its true character as a succession of coastal river deltas was recognized, the distribution and manner of occurrence of its auriferous portions was much clarified. Where it has been crossed by more recent river systems, they have been to a large extent enriched thereby; notably in the case of the Smith River in Del Norte County and that of Cottonwood Creek in Siskiyou County. The fact that in many places thin beds of Cretaceous and Eocene sandstone overly the conglomerates has led to some confusion. This merely proves, however, that intermittent subsidences, taking place after the river deltas had been formed for long periods and the shifting beds of the rivers had distributed their burden of gravel, have covered this coastal area for considerable depths. In many places this conglomerate belt can be found resting directly on Cretaceous deposits, although the early rocks are not far beneath. The presence of well defined rims in a few places indicates a complete submergence of the coastal rivers by this subsidence as the coast line receded.

The general strike of the axis of the Coast Range through Del Norte and Humboldt counties is northwest and southeast, with a southwesterly dip as the crest of the range is nearest to the eastern slope. The Tertiary rocks preponderate over those of the Cretaceous. From Requa northerly to the extreme end of the State, the country is very rugged and covered with forests, the rocks greatly resembling those of the Sierras. The gravels of the rivers carry both gold and platinum values. In this northern region serpentine is the principal rock. Peridotite is also found in places. The greater part of the geological formation of the Smith River country is composed of sedimentary rocks of Tertiary and Cretaceous age. Apparently, though there are granite outcroppings, it does not form the axis of the main Coast Range. In the western portion of this country sedimentary rocks are prevalent, although toward the east granite forms the nucleus of most of the ranges. The slates carry thin seams of quartz which are often rich in gold.

Almost all of the gold-bearing gravels of the Smith River Basin contain black sands which carry some platinum. The beach sands also carry values in gold and platinum.

The Cretaceous shore-line, carrying on from Hayfork, hit the present coast-line somewhere north of Trinidad, turning north by way of Gold Bluff. It passes over the Klamath east of Requa and runs northerly through Del Norte County across the Smith River, crossing it on both the south and middle forks. The black sands of the coast in this region have been a field of investigation for many years, containing platinum and a very fine comminuted gold. Wherever the shore conformation and the direction of wind and tide have been right, there has been a considerable concentration of both gold and platinum. A plant built two miles south of Crescent City on a so-called magnetic repulsion prinicple proved a complete failure. The plant was located just back of the shore-line and consisted of a suction pipe, a conveyor to the plant, a large area of aluminum plates with riffles, a second area of small aluminum plates similar to the first, and a third metal plate of unknown composition, together with electrical equipment for charging the plates. The sand was delivered from the suction pipe line to a revolving screen. The sand from the screen was hauled up an incline to the top of the treatment plant where water was added in sufficient quantity to give it flowing properties, four thousand gallons of water per minute being used. The sand and the water flowed over the riffle area at the same time that an alternating current of electricity was passed through the plates. The black sands were supposed to be repelled, leaving the gold and platinum in a concentrated condition. Theoretically, a good extraction was supposed to be made, but practically and commercially, the plant was a total failure.

At Gold Bluff the steady erosion and concentration of the Cretaceous shore-line resulted in values which in the earlier days permitted carrying the sands out by mule loads for treatment. A shovelling plant was later put up which was unsuccessful, as the thin streak of auriferous sand was covered by too great an overburden of barren material. The same thing occurred at Big Lagoon. The Cretaceous shore-line, crossing by Salt Creek over to Big Lagoon and running north along the coast to Gold Bluff, turns northerly and crosses the Klamath above Requa. From here it runs northeast to Smith River. The region between the south and middle forks of Smith River has large areas of semi-cemented gravel of low tenor, in places concentrated by recent streams. The commercial value is not very great, although deposits cover large areas.

At Harris Flat, between upper and lower Coon mountains, at Big Flat and on the ridge between Hurdy Gurdy Creek and Gordon Creek, the gravel has been prospected with little encouragement. It is situated from 2000 to 3500 feet in elevation and has been prospected to a depth of over 70 feet without reaching bedrock.

At Rattlesnake and Little Rattlesnake mountains a deposit of large acreage exists which has been prospected to a depth of 300 feet. This shows a very low gold content. The gravel seems to be of three different ages, and the shore-line has probably changed at least three different times. Each stratum of gravel has a different character of gold. The only commercial producer of this section was Big Flat. This paid well at one time but the pay gravel has long since been worked out. Big Flat was undoubtedly enriched by the concentration of higher gravels.

Small bars along the north, the middle and the south forks of Smith River have been worked and there is much gravel left that is of doubtful value. The south fork was the richest because its drainage has cut all of the high shore gravels the most frequently. At the junction of the north and middle forks of Smith River there is a large bar with very heavy wash and very light gold content. At the junction of Smith River and Myrtle Creek a good bar was mined in the earlier days but was probably locally enriched, as Myrtle Creek has been the best of any. It was probably fed from local ledges and stringers. As high as $1,100 has been taken out in one piece on Myrtle Creek. Practically all of the bars are well worked out at present. Craig Creek has produced both platinum and gold for many years and some ground still remains on it. Monkey Creek has produced heavy gold and still has some unworked ground on it. Mill Creek and Clark Creek were also worked considerably in the early days. Clark Creek is reputed to have paid as high as $70 per man per day. This is a concentration of the old shore gravels.

The mines of the Smith River were mostly small hydraulic properties where one or two small giants were used during the winter months. In many instances a plain fire hose and nozzle were used on the gravel banks, and the water collected from the gulches in the rainy season was stored in small reservoirs.

This country is accessible mainly from Trinidad northerly along the coast by wagon road and also from Grants Pass by wagon road. There are no railroads at present constructed through the county, and the distributing point at Crescent City is only reached by small steamers and schooners from the south. The principal mines of the district are the Aurora Hydraulic Mine, located at French Hill; the Dr. Young Mine, also located in the French Hill district; the Dave Savoy Mine in the French Hill district; the Elkhorn Mine, located at the mouth of Patrick Creek; the French Hill Mine in the district of the same name; the George Washington, situated on Monkey Creek; the George Cook Mine on the middle fork of Smith River; the Kaus Mine, situated in the Craig Creek mining district; the Myrtle Creek Mine, situated in the district of the same name; the Monkey Creek Mine, situated in the Monkey Creek mining district; the Nels Christensen Mine, situated near the junction of the south and middle forks of Smith River; the Oak Flat Mine in the Patrick Creek district, and the Walter Cook Mine in the French Hill district.

The principal water supply for mining purposes comes from Patrick Creek, Craig Creek, Monkey Creek, the south and middle forks of Smith River, Myrtle Creek, and Shelly Creek.

In general, the gravels of Del Norte County are no longer of great economic importance, their tenor being too low for operation on any large scale. There are, however, possibilities for the small miner content with making moderate wages. The beach sands of Del Norte County present a problem on which hundreds of thousands of dollars have been spent with little successful practical result. It is possible that some means of extracting the gold and platinum, which undoubtedly exist in these deposits, may yet be found, but to the writer's knowledge there is at present no way of making them commercially productive.

Klamath River and its tributaries have been the most important agents in the distribution of placer gold through Siskiyou County. The length of the Klamath is about 362 miles, but only the portion from Hornbrook down to the mouth of the river is of interest to the placer miner. In this section gold is found wherever the river has deposited gravel, whether it be in the old channel several hundred feet above the present stream or in the present river bottom. Below the mouth of the Scott River the Klamath has cut through the northern end of the Coast Range practically at right angles to its trend. This portion is marked by very steep and rugged canyons with occasional flats and high bars, in which the river has meandered in former ages.

Commencing at the mouth of the Klamath River, the first natural division with regard to physical characteristics is from Requa up to Tule Rapids. From Tule Rapids to Weitchpec the river increases its grade, but from Tule Rapids down to the mouth it is marked by a much lighter grade and by relatively smaller gravel. From Weitchpec down the lower Klamath is practically virgin ground so far as the placer miner is concerned. One reason for this is that a large portion of the river is still in the Hoopa Indian Reservation, and has not as yet been opened for prospecting. As stated, above Tule Rapids the bars are not so frequent and the wash is fairly heavy. Below Tule Rapids and running to Requa is an area which still has some promise for prospecting and exploitation by large capital seeking investment in placer mines. The only means of access at present is by motor boat up the Klamath River from Requa.

Extending from the low bars of the river back for several hundred feet above the river are high bars which carry gold to a depth of a hundred feet and over. In most cases water would have to be brought from tributary creeks of the Klamath for several miles. Below the falls at Tule Rapids the wash in the present channel and in the low bars becomes much lighter. At Johnson's Bar there is a long bar with about 50 feet dump into the present river, which was worked for many years to a depth of one hundred feet by the owners, who were Indian allottees, in a very small way by water that was brought from Pequam Creek. This bar is said to have averaged about 25 cents a yard. The wash is very light. There is a high bar above this of about the same depth. From here on up to Tule Rapids bars are frequent on both sides of the river, and of such a nature as to be possible hydraulic ground.

There is much ground below here which seems to have some possibilities for dredging purposes; notably near Johnson's, at Blue Creek, at Blakes, at Terwah, and from Terwah down almost to Requa. The wash appears to be light and there should be some values in the ground which averages around a hundred feet deep. If it contains sufficient value, this ground could best be worked by one large company which could control all of the bars and work directly up the river. The allotments can be purchased by consent of the allottees, together with that of the Indian Agency. This consent could probably only be gained by the transportation and consolidation of all of the local Indian families to a few bars on the river where they could establish a permanent dwelling place. One of the large bars might be set aside for a settlement. If properly cultivated, any one of the larger bars could support the entire population. Graveyards would, of course, have to be respected. This is perhaps an opportunity worth investigating, as there is very much virgin ground of known richness which has only remained unworked by reason of the fact that it was in Indian hands.

Owing to the fact that little hydraulic mining was done in the early days in this section, the bars between Johnson's and Weitchpec are still mostly unprospected. In only three bars on the lower Klamath has any work been done, and records of this work are absolutely lacking. These bars were at Johnson's, at Pequam Creek and below Orcutts ranch. The best water right in this section is from Pequam Creek. Dump is good all the way down. Water can also be obtained from Blue Creek. Since early days this has been Indian territory, and the bars have been alloted to the Indians for use and for their homes, hence, mining has been forbidden, and for this reason there has been very little prospecting done. Areas on these bars should be segregated by drill prospecting, and no attempt should be made to mine all the bars, as there is undoubtedly much of the ground that would not pay.

For hydraulic mining many high channels appear to be left, at elevations approximating a thousand feet above the river. Some of these prospect very well. Water and dump are good and the bars are not heavy.

Most of the lower portion of the river is in Franciscan slate and sandstone. Shortly below Tule Rapids these are replaced by the Paleozoic metamorphics. The bars of the tributary creeks do not carry much gold. It is only in the old channels of the Klamath that good values appear to be found; although there are no mines operating in this vicinity. From Tule Rapids to Weitchpec the Paleozoic rocks are predominant. Weitchpec is at the junction of the Trinity River with the Klamath. From here on up the grade becomes steeper, and the bars are less frequent. The water supply is less from Weitchpec down to Tule Rapids than it is from Tule Rapids down to the mouth.

At Weitchpec there is a long high bar on the Klamath on the Indian Reservation, which might pay to work by hydraulic mining. It has an excellent dump and water could be brought from Bluff Creek or from Hopkins Creek on the south side of the river. Below Hopkins Creek is a plaster bar with the front rim gone, and a very heavy top, which has been worked unsuccessfully. At Bluff Creek and above at Big Bar a high bar was worked unsuccessfully.

It is reported that the saving was very poor and that much gold was run over. At two or three small bars on both sides of the river in this vicinity mining is still being carried on in a small way. At French Bar much money was taken out in the early days. Below this, at Red Cap Creek on the south side, is a bar which might pay to work from its favorable location, good dump and excellent water right. It is about fifty acres in extent. Below Camp Creek are low bars which look good but have poor dump. The Salstrom Place and the Wilder Ranch have a large area of workable ground which should be good if properly prospected first, in order to direct the mining work intelligently. At Orleans Bar is a large area of unworked ground on both sides of the river. On the north side the front ground has been worked off. On the south side of the river the old Perch Mine was the best paying ground and considerable ground is left; but the bars of the Orleans

Company, both the high and the low, contain a much larger area of gravel, averaging about ten cents, with a very good dump. The water is brought from Camp Creek. There are some bars on Red Cap Creek and also on Camp Creek which have paid in a small way. Outside of this, few of the creek bars have paid to work. This country is almost altogether in the Paleozoic metamorphics. The grade is much steeper and the boulders are heavier. On the other hand, there is much more water for working purposes to be taken from Hopkins Creek, Bluff Creek, Camp Creek, Slate Creek and Red Cap.

The lower portion of the river is only accessible by trail or by boat, but the region from Weitchpec up is accessible by wagon road for about seven months of the year from the coast. Most of the ground above Weitchpec would only be suitable for hydraulicking, on account of the size of the boulders and the character of the bedrock.

Above Orleans are a few small bars between there and Somes. These are mostly unimportant, but some of them have been worked in a small way. The Bondo Bar was worked in the early days and paid very well. The Reese Ranch and Fish Ike's place are bars of fairly good extent. The Perch property, now belonging to Mr. Young, still contains much good ground in the form of a high bar. The Nelson property contains at least a mile of the old channel of the Klamath River with two or three benches on it. It is said to prospect well in places. It has a dump of from 500 to 600 feet, but the water problem is almost unsolvable. Water would have to be brought all the way from Rock Creek or Dillon Creek, as both the Reynolds and the Ten Eyck Creek water rights are not very large. The installation would certainly be very expensive. The Hickok Mine on and near Ten Eyck Creek extends about half a mile along the river. Here there is very heavy wash and a very good dump. There are several small bars on the south side of the river above here all the way to Farnums Ranch. The Sphinx Bar opposite, has a low and a high bar on a short turn of the river. At Harley's and at the Lord Mine on the opposite side of the river is a short turn, with at least two courses of the Klamath upon it. Much gravel is left at the Lord Mine, but it apparently could not have been pay. From Harley's Ranch up the river for two miles are numerous large, high bars with considerably heavy wash. There is excellent dump but not much water. On the north side of the river, below Rock Creek and extending out above Dillon Creek, is an almost unbroken bar which has been slightly worked on the front rim. The water right and dump are excellent. At Blue Nose and one other bar below, there is considerable gravel that is reported as of only moderate value, although it has been worked for many years. At Thomas Ranch and at Aubery's, as well as at Elliotts and Cottage Grove, a good deal of gravel is still left as both high and low bars. Most of the above mentioned gravel will run around ten cents a yard, and would have to be handled on a large scale in order to make any profit. There are spots where there is richer ground. This country is practically all in the Paleozoic metamorphies. Most of the hydraulic mines of the Klamath River were in this section. It is now accessible by road. Somes Bar on the Salmon River, near its junction with the Klamath, is the center of a gravel region which will be discussed under the head of Salmon River. The country from Orleans to Somes is a pocket and stringer country in which many small quartz mines have operated. Prospect Hill above Orleans has contributed much by its erosion to the richness of the bars of the Klamath River. The principal water rights in this section are from Rock Creek and from Dillon Creek.

Immediately above Elliott's is a low bar and then a high one. About three miles above is a long low bar. From here on there is nothing until Crawford's Creek is reached. Beyond this there is a series of high and low bars: some of them well worked and of considerable extent, until about nine miles below Happy Camp. The Siskiyou Mining Company worked a large area in this section. Most of the gravel has a good dump and fairly heavy wash. A high bar extends on the east bank of the river for several miles below Happy Camp, which has only been worked intermittently. It has a dump of 300 to 400 feet. Two miles below Happy Camp and extending from six to eight miles above is a series of high and low bars of considerable extent. The Davis or Van Brunt Mine and the Siskiyou Mining Company are the principal companies that have operated here. On the low bars, if the gravel were prospected sufficiently, there is an excellent chance to use a Ruble grizzly here, as there are many long, low flats with little dump which might pay to work. At Happy Camp and south, on both sides of Indian Creek, is a low bar which might be worth prospecting. There is about ten feet of gravel, but it is said to be very good on bedrock. There is little dump here. About five miles above Happy Camp there is an opportunity to turn the Klamath River by one and a quarter miles of tunnel and secure sixty feet of drop for power purposes. Power for driving this tunnel could be furnished by the flume of the Siskiyou Mining Company. This will leave nearly seven miles of the Klamath channel dry for mining in the summer. As this portion of the river has never been wing-dammed, it might contain considerable gold. The creeks in this neighborhood have not been worked much. Indian Creek, which comes into the river near Happy Camp, was very spotty. The gravel has a low content, but pocket mines and hill sluicing operations, such as at Classic Hill, are reported to have been very good. Below Happy Camp the principal water rights are from Clear Creek and Crawford Creek.

Titus Creek, Elk Creek and Independence Creek, all heading on the Marble Mountain side, have been worked to a considerable extent, but there are still numerous small bars on these creeks which might be attractive to the small miner and prospector. In this region we pass into the older metamorphic belt which has been classified as pre-Cambrian. From Happy Camp to Hamburg Bar these rocks are predominant. The principal enrichment of the placers in this region must therefore have come from the erosion of the gravels of the second cycle, which have already been mentioned in the first part of this chapter. In the neighborhood of Portuguese Creek traces can be found of the crossing of the old northward flowing channel of the Cretaceous island which had its outlet in Oregon on Galice Creek. The principal water rights of this section come from Grider Creek and Thompson Creek, as well as Seiad.

Six miles from Happy Camp by road and about eleven miles on the course of the river gravel bars begin again at or near the Woods Mine below Thompson Creek. From here on the gravel is intermittent with very heavy wash all the way to Seiad. The Seiad Valley appears to be a possible dredging area of several hundred acres in extent. Immediately above Seiad is a hydraulic bar which has not yet been worked out. In fact, from Seiad up to Hamburg is an almost unbroken section of bars of fairly good dump. There are high and low bars on both sides of the Klamath, most of which have been partly worked, while others are virgin. The water here is a rather difficult problem, although Grider Creek is the best water right. This country is all accessible by wagon road.

From Hamburg up to Oak Bar, still passing through the zone of pre-Cambrian metamorphics, we find most of the bars to be barren. There are small bars on Horse Creek which have been unprofitable; and on the Scott River, which takes its entire course through the later metamorphics, there is an exceptional enrichment, which is probably the cause of what gold content there is, in the bars at Hamburg and below. On Horse Creek there is considerable gravel of doubtful value, which is in the creek bed and has very poor dump. The higher bars have been worked off wherever there is pay enough to work. Above Oak Bar the river again runs through the Paleozoic metamorphics, and we find considerable enrichment again. The tributaries of the Klamath, among which are Beaver Creek, Humbug Creek, Barkhouse, McKinney and Little Humbug Creek, are all very rich, and still contain some gravel which should be of interest from the hydraulic standpoint. On Beaver Creek is an extensive area of gravel which looks promising. From Oak Bar to Hornbrook the country has been characterized by heavy pocket gold. There are small hydraulic mines on both sides of the river but the wash is very heavy. This region was worked in the early days and, as it was quite accessible, has been fairly well worked out, with the exceptions noted. Near Hornbrook the old Cretaceous shore line is crossed by the Klamath, and this, together with Cottonwood Creek, forms the uppermost zone of enrichment of the Klamath River in the State of California.

The approximate production of the area from Hornbrook down to Hamburg Bar has been over $400,000. The principal mines of the Klamath River were worked in the early days. At present there are probably not more than half a dozen operations. Below Thompson Creek at the Woods Mine; at and in the neighborhood of Happy Camp; in the neighborhood of Orleans; and also on the upper river above Oak Bar, these few operations comprise the present production areas of the Klamath River.

So far as the tributaries of the Klamath are concerned, the Trinity River is the most important; next to this the Scott River, and next to that the Salmon. These rivers will be taken up in the order of their relative importance commencing with the Trinity River.

TRINITY RIVER.

The main Trinity River rises in the neighborhood of Scott Mountain, flows south for about sixty miles, and then makes a detour westerly for another sixty miles until it unites with Klamath River at Weitchpec in Humboldt County. The south fork, which joins it at Salyer, flows in a northwesterly direction from the Yolo Bolo Mountains to the junction. The main river for the most part has its course through the

Paleozoic sedimentaries which are the source of most of the gold in California. On the other hand, the south fork has most of its drainage in the pre-Cambrian metamorphics, which, as a whole, are barren of gold-producing ledges and stringers. Commencing at the junction of the Trinity with the Klamath; for about two miles above, the gorge is very steep. At the Bull Ranch there is a high bar which might be suitable for hydraulic mining, although water is comparatively inaccessible. Above this, the Hoopa Valley, which is one to two miles in width and six or seven miles in length, contains many flats and low bars which probably have dredgeable areas within them. The gravel, however, is shallow and could only be worked by a small dredge or a Ruble grizzly, as the dump into the river is very poor. The gravel is not very heavy and the water rights in the Hoopa Valley are good, the best one being from Mill Creek. Most of the land, though, is allotted to the Indians and is their sole support. For this reason it would be difficult to obtain permission from the Government to mine these bars. There are about three hundred to five hundred acres of unallotted land that is suitable for placer mining. Some of this ground is reputed to contain very good values.

Above the Hoopa Valley, the Sugar Bowl Ranch offers the next possible dredging area. There is a tremendous amount of high wash gravel on the mountain between Hoopa and Willow Creek, but the gravel is very low-grade and water is inaccessible. Below Willow Creek, for three or four miles, there are wide flats which, with proper prospecting, might segregate areas of dredging or Ruble ground.

From Willow Creek up to Salyer are several high bars which might contain portions suitable for hydraulicking. The dump is good and the wash is very light. The water problem is a little difficult however. There are several large bars at the junction of the south fork at Salyer. These bars are both high and low.

The south fork of the Trinity is a stream along which little mining has been done. From Auto Rest down to Trinity many bars occur, which might have possible hydraulic ground among them. Some of these bars contain a large acreage. At Hyampom there is a large acreage which might prove good dredging ground if there is sufficient value and depth. It probably contains platinum, and might pay to prospect for this metal. At the head of Corral Creek some heavy gold has been taken out from the old river wash which occurs there. This is on the divide between the south fork and the main Trinity River. The pay is fairly good for ground sluicing work, and there seems to be some ground left which might pay to work. Between Hayfork and Hyampom, along Hayfork Creek, a little mining has been done but not to any great extent. A serpentine belt crosses the south fork, and fairly good platinum prospects have been found at some of the bars. It is possible that platinum might be the main valuable constituent of these gravels. Ettapom Creek has given results in platinum prospecting but has not been mined very extensively.

The south fork, as a whole, is very rugged, and water installations would be very expensive. The watershed is precipitous and the water supply is naturally limited in spring and summer. Grouse Creek has the best water right of any of the tributaries. Most of the gravel is rather fine wash, as heavy wash only crosses at the foot of the canyons and falls. Most of the bars have good dump, but the expense of bringing on water would, in many cases, be prohibitive.

At the head of Brown's Creek and on Duncan Creek, extending down to Carr Creek, is a high channel similar to the La Grange, and it might possibly be the same channel emptying into the shore line which, crossing below Hayfork on Salt Creek, runs over below Gold Bluff in Humboldt County. This channel can be traced down Hayfork for several miles. It has the same type of dead wash in it as the La Grange, and the gravel is not very heavy. There is a large area in the Hayfork Valley which looks as if dredgeable areas might be selected in it. Near Wildwood on the Hayfork is considerable platinum.

Continuing up the main river from Salyer to Burnt Ranch, the main river is interesting for its platinum content. There is an almost unbroken succession of high and low bars, mostly on the south side. At Burnt Ranch itself there is probably an excellent hydraulic property if portions of it were properly segregated by drilling. Water is available and the dump is excellent.

Above Burnt Ranch at Cedar Flat is good mining ground on both the high and the low bars. A short turn of the river at Don Juan should have some good ground on it. Some good ground is still left at Taylor's Flat and at French Creek Bar, but without very much dump. At Big Bar itself is an inside channel two miles long that has paid well on the upper end at the Tinsley workings.

Below Burnt Ranch is the junction of the New River with the Trinity. At the mouth of the New River there appears to be a small gravel bar of about ten acres that prospects fairly well, but it is not of sufficient extent to warrant putting water on the claim. The first real gravel bar as we ascend New River is the Siegler. On both sides of the river is a large deposit of gravel of about 150 acres. It has never been prospected and is a very likely looking bar. Water could be obtained from Big Creek on the south side and from Bell Creek on the north side of the river. Bell Creek is about a quarter of a mile above the ranch, but Big Creek is the better water right. Above Siegler's is Hoboken Bar, also called Grant's Slide. It has been mined with water from Bell Creek, but no record as to the values is obtainable. A large bar is still left and might be worth prospecting. Big Creek on the opposite side is half a mile below this bar. The next bar above is a small claim at the mouth of China Creek, about twenty acres in extent. It has not been mined. A bar immediately above this has been totally mined off, however. This bar contained very heavy gold, like most of the New River country.

Above China Creek another large bar is exposed which contains over a hundred acres of gravel. A company started operations ten years ago with water out of Bell Creek and quit, but no definite reason could be ascertained. A good water right from Panther Creek could be put on this claim. Above Noble's Bar on Panther Creek is Henderson's Bar and the Nigger Mine, which is being worked at present. The Schoolhouse Bar is another portion of this property. Above this, the Burchoff claim has been mined on the rim, but the main bar is left. Above this, Jackass Bar was mined in the early days and is claimed to have been good pay. The Owens Bar, above this, has been mined with water from small gulches, with no record as to pay.

The New River Mining Company used to own a very large bar on the south side of the river above the Owens Bar. This property is now owned by the Ladd Brothers. Mining has been done on this bar for many years, and it is claimed to be good if properly prospected and segregated. A good water right from Quimby Creek is on the mine, consisting of about 2500 inches under 150 foot head for about five months out of the year. Approximately a quarter section of ground is left. A couple of high bars are also above this claim. This is about the biggest property on New River. Across from this is the McAtee claim, which is a short turn of the river with three separate channels. Some of this ground should be good. It paid 30 cents a yard when mined during 1920 and 1921. Water is piped across the river from the Ladd claim.

At the mouth of Devil's Canyon is a fair sized bar. It has not much dump, but the values are said to be good. Water from Devil's Canyon could be brought to all of the bars in the vicinity. This is a very good water right. Pony Creek, which is a tributary of the east fork, is reported to have produced about $3,000,000 in the early days. It has practically all been mined off. The east fork and New River have been wing-dammed by Chinamen with good results.

The Slide Creek channel is supposed to contain some good ground, but the dump is poor and the wash is quite heavy. Some of the ground was prospected by the Tener Mining Company, and their results were 36 cents a yard for about 22 feet in width and 30 feet in depth. The Boomer Quartz Mine at the head of this creek is reported to have produced about a quarter of a million.

Eagle Creek and Virgin Creek should be interesting to the small miner, as they carry heavy gold in rather small bars and flats. Emigrant Creek was good, but is mostly mined out. Quimby Creek was also a good producer.

Coming back to the main river, for about four miles above Big Bar, are small placer bars either on the river or above it, which are all courses of the present Trinity River. Some of the gravel is quite heavy. Many of these bars have a production record and still have some good gravel left. From this point on to Helena the canyon was too steep to permit the forming of bars. Above the north fork, Red Hill is still a good hydraulic property if water could be brought to it.

On the north fork itself are some good flats in which small dredging areas might be segregated. Rattlesnake Creek, Grizzly Creek and White's Creek still have some good bars in them, which might be attractive to the small miner. Both the north and the east forks have considerable ground on them that might be suitable for a Ruble Grizzly. On the east fork of Yellow Jacket Creek both Rich Gulch and Crump Gulch were very rich and still have some good ground left on them that might be attractive to the small miner.

Canyon Creek has been worked for fifty years and still has much good ground left on it for hydraulic purposes. The best chances for a large mine lies seven miles up this creek. The dump is excellent, but water would have to be brought six miles to work the high bars. Five acres that were mined off are reported to have averaged about 40 cents a yard. A ditch line from the east fork would bring 2000 inches or more of water for six months. The bedrock is slate and the ground very easily mined. The bank averages from 30 to 60 feet deep and has some fine loam on top.

North Fork, now named Helena, has one or two high bars that were worked in the early days, and some ground is still left. From North Fork up to Lewiston are many low bars on which good dredging areas can be segregated. The Valdor dredge is operating above Helena, and the Gardella dredge is operating on the Paulsen Ranch, near Lewiston. The characteristic of the dredging ground on the Trinity River is that about 20 to 40 per cent of the values are in heavy gold which, when fine screens are used in the trommels, passes through them and out on the stackers to be lost with the boulders.

In this country much hydraulic ground is available in high bars, provided water could be brought to it. The area of most promise is above and below Douglas City. At Steiner Flat, at Douglas City, Dutch Creek, Brown's Creek, Grass Valley Creek, Redding and Indian Creeks are large bars of gravel in the form of high bars, as well as some low ground, which is practically virgin and presents an excellent opportunity for possible hydraulic ground.

Weaver Creek itself has been well worked, but some gold is still in the tailings which fill the bars. Some good high ground is still left at the Union Hill Mine. From Douglas City up to Lewiston are many high bars on the Trinity that appear suitable for hydraulicking, as well as low bars that may possibly be dredging ground. A large amount of possible dredging ground is in the vicinity of Lewiston, in which profitable areas can probably be segregated.

The La Grange Mine at the head of Oregon Gulch, which has for years been known as the largest operating hydraulic mine in the world, has been closed down since the War. A large quantity of low-grade gravel area is still left, but the cost of opening it up again would be considerable on account of the necessity of a tunnel and deep cut. The water right on the La Grange Mine came from the east fork of Stewart Fork for a distance of about 30 miles, and is one of the finest hydraulic mining rights in the State of California. If this ditch could be tapped about seven miles back and above the La Grange Mine, there is an area of gravel on Musser Hill that would possibly make a larger and better paying mine than La Grange. This gravel was deposited by the same southward flowing Cretaceous channel as La Grange and prospects on the surface even better than La Grange did. Dump for this ground would have to be purchased on Brown's Creek.

A large amount of gravel is still unworked in the basins of east and west Weaverville Creeks, some of which appears to be very attractive. The old-time miners, working on the shallowest banks, covered much good ground with their tailings, but there are several hundred acres, both of flat and bench ground, which are still virgin. The gravel is not coarse, and in many cases is subangular. Most of this gravel is derived from the erosion and concentration of the old Cretaceous channel which runs through from Brown's Hill and Musser Hill to La Grange. Above Dutton Creek is some ground from this channel.

Under the town of Weaverville is a large amount of good gravel which has not been drifted out. The whole basin is largely controlled by the Union Hill water right, but the upper half of it is undoubtedly hydraulic mining ground with excellent possibilities. The Lorenz

Brothers are now working a large and successful hydraulic mine on the lower end of this basin and stacking their tailings. The boulders are not very large, and the grade of the channel is light.

Returning to the main Trinity River, from Lewiston to Carrville, are flats which contain a considerable acreage of probable dredging ground which may be segregated. It is to be hoped, however, that this ground will be worked with flume dredges instead of those of the stacker type, not only for the reason that the ground can be more readily resoiled, but because, unless large screens are used on the stacker dredges, nearly one-half of the gold is apt to be carried through and redeposited with the tailing piles. In the writer's opinion, the Trinity Gold Dredging Company has the only dredge on the Trinity River that is saving its bedrock gold by digging slowly and carefully and putting everything through its sluices; also, it is leaving the ground level and in such shape that it can be reclaimed easily for farming by damming the river and overflowing it.

Deadwood Creek has been worked a great many times by small placer miners, and has proved very rich, but there appear to be isolated patches of gravel remaining that might be suitable for the small ground sluicer or drifter. The same thing may be said of Jennings Gulch and Eastman's Gulch. On Eastman's Gulch is a considerable body of possible hydraulic mining ground.

From Carrville on the west bank of the Trinity down through Minersville and Buckeye Ridge, Brown's Hill and through Weaver Basin extends the southward flowing Cretaceous channel. This was greatly eroded in the Weaver Basin, and every creek and gulch that cut this channel was enriched. Much of the channel has been worked near Trinity Center, but above Minersville on the east fork of the Stewart Fork are several miles of virgin ground which should make possible hydraulic property. The Beaudry, the Unity and other properties are on this channel. Near the junction of the east fork and the Stewart Fork is some possible dredging ground. This channel is the same as that which goes through Musser Hill and La Grange and on to the Hayfork country. To the north it has been practically all eroded above Trinity Center. It can be found on both sides of Coffee Creek and also on Scott Mountain.

On the east fork of the Trinity River a large area of possible dredging ground still remains. There is a tributary high bar on Plummer Hill and on the side of Paul's Gulch. There is also a more recent channel on Crow Creek running over to Slate Creek, and also on the head of Snow Gulch is considerable recent gravel.

Coffee Creek has been mined for many years. The upper portion of it was evidently drained by what is now the south fork of the Salmon River in preglacial times. A few small placer mines are still on Coffee Creek, which may have some value. The Nash Mine, later known as the Big Flat, is one of these. It has had a fair production record and undoubtedly has some good ground left.

Dredging operations in Trinity county have been largely confined to the middle fork of the Trinity River. The largest dredges are operating in the section between Carrville and Trinity Center. Below this, a flume dredge is operating above Lewiston; another near the Rush Creek confluence; and another above Helena.

SALMON RIVER.

The southwestern portion of Siskiyou County comprises the drainage area of the Salmon River and its tributaries. The topography is very steep and water installations quite expensive. This district is connected by a wagon road forty-three miles in length between Etna and Scott Valley and Forks of the Salmon, located at the junction of the north and south forks of the Salmon River. As most of the level ground in the district consists of the low lying bars on the Salmon River and its branches, this section is practically dependent for food and supplies upon communication with the outside which is only open for wagon or truck traffic for about five months out of the year.

In the early days the Salmon River was famous for its front rim diggings in low bars close to the present river. These diggings were very rich. In the main, since the days between 1850 and 1870, the larger hydraulic operations have been unsuccessful. At present very little

PHOTO No. 19. Scott Mountain Region, Siskiyou County, California.

mining is being done on the river, as the best of the available ground has been worked out. The Salmon River suffers from many drawbacks. In most cases water installation is very difficult and expensive. The top is heavy or else the dump is very poor, and many heavy boulders hamper the work. The pay is generally confined to a narrow area and the greater portion even of some of the best bars is too low grade to be profitable. Commencing at the upper end of the north fork, the best placer remaining lies in White's Gulch at the Craig Mine, and in Eddy's Gulch at the Peterson Mine, both of these gulches being tributaries of the north fork above Sawyer's Bar. Due to the erosion of the old Klamath ledge and other blanket ledges, Eddy's Gulch and Sawyer's Bar, just below it, were greatly enriched. With the exceptions named above, most of this country is now worked out. Above Eddy's Gulch on the main river there is considerable gravel in high bars, but as a rule it has not paid to work. From Sawyer's Bar on down are still hundreds of acres of unworked high bars with very heavy top, heavy boulders and good dump. One or two small mines are working in desultory fashion, but without enough water to accomplish much. Below Bonaly Bar, which is about three miles above Forks of Salmon, is some good ground and much gravel is left, some of which is said to be pay gravel. At the Forks itself a fair sized flat is still left which contains some good ground, but this has been largely drifted. The only water that is available for mining this ground is that formerly used by the Forks of Salmon Mining Company. This flume could be repaired and extended so as to cover the ground at Forks of Salmon.

Up the south fork of the Salmon considerable work has been done as far up as Orcutt's Ranch, and some small bars are still left which probably would not pay to work. The principal gravel areas of the south fork are at Cecilville and Summerville and extend clear up to the Coffee Creek Divide. There is a tremendous amount of gravel here, but the pay is very irregular and spotty and, as a rule, mining opera-

PHOTO No. 20.   Scott Mountain Region, Siskiyou County, California.

tions on a large scale have never paid and probably never will. Two small mines are still working near Cecilville.

From Forks of Salmon down to Somes Bar is much gravel, but very little of it has paid to operate. Codfish Hill, the Bloomer Mine and the Nordheimer Flat have been the best mines on the river, but in all of these the greater proportion of the pay gravel has been worked out. Only two mines are working at present in this section of the river. The Bloomer Mine is entirely different from any other channel on the river, and in the nature of its wash and the type of its gold this more closely resembles the northward flowing Cretaceous stream than any other. If it is a portion of this stream, it has probably been almost entirely eroded for many miles. It might possibly have flowed northward toward Portuguese Creek between Happy Camp and Hamburg Bar.

A few opportunities for the small miner without an expensive installation still remain at and near Portuguese Bar and on the north fork below Sawyer's Bar, as well as above Cecilville on the south fork. A little good ground is also left on Nordheimer Creek.

SCOTT AND SHASTA RIVERS.

The last three miles of the Scott River, before its junction with the Klamath, are the richest portion. Practically all of this distance has been wingdammed wherever possible. Some low bars are still left, but outside of the flat on which the town of Scotts Bar stands, they have been drifted out. A deep channel is still left in this vicinity which might pay if the water is not too hard to handle. A great proportion of the enrichment in this part of the channel comes from the erosion of Quartz Hill. This hill has been hydraulicked for many years, and is still being operated at a slight profit. About a half mile above the junction of the Scott and Klamath rivers is a high bar at the Roxbury Mine, which is probably one of the most interesting hydraulic propositions in California today. There are in the neighborhood of five to ten million yards of gravel, which will probably run from ten to fifteen cents a yard. The present water right is not sufficient for the profitable working of the property. It will be necessary to bring water from Thompson Creek, and about twelve miles of the necessary ditch is still uncompleted. Considerable money was spent here in 1917 by the present owners but not to good advantage. The grade used on the sluice boxes is entirely too high for the saving of any fine gold with the amount of water that is used.

Continuing easterly, the principal gold-bearing tributaries of the Scott and Shasta rivers were Greenhorn, McAdams and Cherry creeks. A deep channel with cemented gravel, which is probably a portion of the delta gravels of the Cretaceous island before mentioned, underlies Greenhorn Creek. There is a tremendous amount of water in this, and it is now used as an auxiliary water supply for Yreka. A large proportion of the gold-bearing gravels of Greenhorn Creek has been partly dredged. McAdams Creek has been dredged for the greater part of its length, but on the lower two or three miles there is still some virgin ground which might possibly be capable of yielding a profit. Some hydraulicking has been done on Greenhorn and Cherry creeks, and some good ground still remains.

Quartz Valley, which is tributary to the Scott River, has the appearance of a possible dredging proposition, which has been fed from the gulches on either side of Evans Creek. Considerable good ground still remains in these gulches for the small operator, notably on the east side.

A large channel, which is possibly the old Tertiary equivalent to Scott River, which contains subangular and slightly cemented wash, comes down from near Callahans through Etna and across the upper end of Quartz Valley, and into the present bed of the Scott River where it turns westerly toward Scotts Bar. Spills from this channel enriched the old Etna Creek diggings and faulted portions of it were worked at the Old Piney and the New Piney diggings on both sides of the ridge west and south of Evans Creek. This channel has the possibility of being one of the largest hydraulic propositions yet remaining in the northern end of the state, but considerable ground would have to be bought for dump purposes and water would have to come from the Scott River.

Oro Fino Valley appears to have some possible dredging ground. For a great distance on both sides of the valley the slopes of the ridges have been worked since the early days. Fed from a rich pocket country a great mass of subangular erosional detritus still occupies the center of this valley, and evidently contains some gold. Considerable areas of ground, possibly suitable for hydraulicking, still remain on the ridges. In all cases where this has been mined, the bedrock has evidently been very carefully stripped and cleaned. Presumably most of the pay lies close to the bottom.

Below the junction of the south and east forks of Scott River, north of Callahan, is still some excellent dredging ground in spite of the fact that one disastrous and abortive attempt was made to handle it several years ago. Above this junction of both the south and the east forks there is also some promising ground. In this region one of the best areas of hydraulic ground also remains, which is suitable for the small operator with little capital. Unfortunately a large proportion of this ground is tied up in the Beaudry Estate, lying on Wildcat Creek and also on the south fork. However, on Fox, Slide, Grouse and Kangaroo creeks a small quantity of spotty pay gravel still remains in small segregated areas, which might be interesting to the pick and shovel miner. East of Kangaroo Creek very little pay is encountered. North of the junction of the Shasta River with the Klamath, Cottonwood Creek forms a dividing line in northern Siskiyou County between the auriferous and barren gravels which lie respectively on the west and east sides of it. South of Hornbrook the old blue shore gravels of the Cretaceous island are being worked by drifting at the Bradley Mine. This shore gravel extends south under Yreka and the Shasta Valley, and has been worked at the Blue Gravel Mine west of Greenhorn Creek. Greenhorn Creek itself has been worked by dredging with fair success and on the upper end there is still a considerable area of good gravel. The bars of Humbug Creek and the bars of the Klamath have been worked clear through as far as Hornbrook. Not very much remains on Humbug Creek but a few small deposits near the head. This creek is tributary to the Klamath. Most of the gravel is subangular.

Wherever any gulches have cut and concentrated the old shore gravels the work has been profitable on both sides of the Klamath.

At Hawkinsville a large area of shallow gravel has been worked off by hand. The gravel is from one to ten feet deep and a great deal of it still remains, but water is not available for hydraulicking. It is possible that by using an excavator of the Hadsell type, some of this ground could be handled with a small amount of water.

On Yreka Creek, south of Hawkinsville, some ground still remains which appears to have possibilities of dredging. This lies north of the old dredge workings on this creek. The town of Yreka appears to be located on possible dredging ground, but in order to handle any of the area in this vicinity, a resoiling dredge would probably be required, as the present holders of the land in this locality do not wish to see it ruined in the same way that it has been near Oroville.

Further down the Shasta Valley and north of Scarface Gulch, and running down the ridge to Granada, is an old channel with some well-rounded wash. This contains gold and platinum and where it has not been eroded might possibly be hydraulicked. Water would, however, have to be brought from the Trinity Slope to cover it.

SACRAMENTO AND PIT RIVERS.

The Sacramento River rises in the mountains above the northern boundary of Shasta County, together with the McCloud, its principal tributary. The Pit River has its source in Modoc County. The Sacramento flows southward through the western half of Shasta County in a deep canyon. The Pit crosses the axis of the Sierra Range and joins the Sacramento in the midst of the copper belt of Shasta County. The McCloud discharges into the Pit above its confluence with the Sacramento.

South of Dunsmuir and extending to Redding there is a tremendous amount of gravel on the Sacramento River, both creek gravel and ancient shore wash. It contains little value and is not of great economic importance. Immediately above Redding these gravels are being dredged with indifferent success. The McCloud and Pit river gravels are of no economic importance, although of considerable extent. The lava-capped channel is exposed on the Pit River near Fall River Mills, but to date it has never been worked.

At Portuguese Flat and near La Moine the low bars of the Sacramento were once worked, but with the exception of the front rims did not pay very well. In general, these gravels may be dismissed as of no economic importance.

In the eastern portion of Shasta County the drainage of Deer, Mill and Battle creeks is practically covered by lava for hundreds of feet in thickness. For this reason the gravels are nearly all volcanic pebbles. They are purely Quaternary channels, and no gold has ever been found in commercial quantities in this region. The same condition prevails down Payne's Creek to Red Bluff.

Going north to Redding and out on the Alturas road we come to Montgomery Creek Ranch. A little successful placer work was done on Montgomery Creek near here, but as a rule the gravels are practically barren. At Hayden Hill, near the northern part of Lassen County, is a channel of intervolcanic gravel some two miles in length which has never been productive. Beyond Alturas, on the east side of the Warner Range, is a channel near Cedarville which has been prospected slightly but without satisfactory results. This, also, is volcanic wash. It is reported that a deep blue channel under the shoulder of Eagle Peak in this range contains some auriferous gravel. Coming down from Alturas by way of the Madeline Plains no gravel is noted, the whole country being covered by lava and scoria. Until Long Valley is reached no extensive gravel deposits are noted; and these are Quaternary and of no commercial value.

The western portion of Shasta County is about the only auriferous drainage of this part of the Sacramento River. Just below French Gulch, and including the town of French Gulch, considerable ground on Clear Creek might be worked by hydraulic means. It is not heavy ground, but lies rather low. Four miles above on Drunken Gulch and Shirttail Gulch is a mile and a half of channel which might have hydraulic mining possibilities. On Clear Creek, successful dredging has been prosecuted; also on Cottonwood Creek. Some dredging has been done near Redding.

The drainage on the south shore of the Cretaceous island was not altogether confined to La Grange channel, which probably entered the shore line below Hayfork. This shore line can be traced from Redding southerly and westerly below Centerville to Igo and Ono around Nigger Hill and westerly. Clear Creek, entering the ancient sea near Igo, deposited much rich gravel in a sort of delta. The old Piety Hill Mine still contains some good hydraulic ground. The Gardella dredges are operating on Clear Creek gravel at a good profit. There are three distinct channels at Igo, in all of which it should be possible to segregate some areas of dredgeable ground, probably about six hundred acres in extent. In addition to this, between these channels, are about 2500 acres which may be good hydraulic ground.

Oregon Gulch produced a concentration of shore gravels near Centerville. This was very rich with much shallow rocker ground. The channel heading up on Arbuckle Gulch produced the same condition at Nigger Hill and at Ono. Near the shore line much of this ground is cemented.

Cottonwood Creek in Shasta County, particularly the north fork, was worked on its front rims for the enrichment produced from these ancient shore gravels. The dredger that is at present operating near Gas Point is making a fair platinum recovery, which is probably caused by the erosion of the Beegum channel on the Bald Hills into Cottonwood Creek. This channel was eroded by the middle fork of Cottonwood Creek.

From the western slope of the Yolo Bolo Mountains down Beegum Creek, and extending along the Bald Hills toward Gas Point, this channel can be traced. It carries some gold but more platinum and iridium. This channel has enriched the present Beegum Creek wherever it has crossed it.

At Harrison Gulch some local enrichment has caused the formation of good, small placer diggings below the town. There is no ancient channel in this vicinity, however.

SECTION 2.

FEATHER RIVER REGION.

For the purpose of this report and as a matter of convenience, the gravels in the drainage of Butte Creek and of Dry Creek will be included with those of the Feather River region. The drainage of all three of these streams in their lower courses is through bench gravels and undifferentiated Quaternary deposits, and through the sands, tuffs and clays of the Ione formation higher up. However, the underlying amphibolites come to the surface and it is in this, as well as in the Paleozoic metamorphics that the greater portion of the stream enrichment has occurred. The area west of Big Butte Creek is largely covered with volcanics of Tertiary age, and it is not until the shore gravels around Centerville are reached that we find any notable concentration of gold. Big Butte Creek and Little Butte Creek have both been worked since the early days of California, and several millions have been taken out from them. These creeks have concentrated the gold from two well-defined channels which entered the shore of the old

Cretaceous sea near Centerville. The lower portion of the channels, due to the delta condition, is very spotty, and the pay is not concentrated. Going up from Centerville on the east side of Big Butte Creek the channel down the Nimshew Ridge is of considerable importance from the standpoint of possible drifting ground. This channel extends from Nimshew through Hupps Mill, Powelton, Inskip and Chaparral, one branch of it heading up near Mountain Meadows. This branch has been worked at the head of Chips Creek and Yellow Creek, and can be traced over across the head of Chips Creek through Lotts Mine. Por-

PHOTO No. 21. Hupp Mine on Nimshew Ridge.

tions of this mine were rich and there is still some good gravel in it. Coming down above the Philbrook Valley, this channel was worked at the Carr and Princess mines. Below Powelton it crosses a wide porphyry dyke which is evidently mineralized and contains stringers and benches of pocket gold. Undoubtedly this is the source of the principal enrichment of this channel. Considerable dredging has been done and there is still some operating on the lower reaches of Butte Creek, whose gold is mostly derived from this. The age of the main channel is probably Cretaceous, and all the way down from Powelton through the Nimshew Ridge a large portion of it is still intact. A later channel of Tertiary age crossed and recrossed it many times and this channel, which reconcentrated the gold from the older one, was in its turn cut down by Big Butte Creek, and was the source of most of the enrichment of that stream. Upon this channel three or four successful drift mines have been operated, but as they generally follow the smaller channel in from the exposures on the west side of Nimshew Ridge and stay close to the west rim until they come out again, the fact that the larger and older channel existed was only made evident at the Emma Mine. There is undoubtedly a chance for a large drift mine here, which is well worth investigating.

Another branch of this same channel is known as the Magalia Channel. The junction of the two was probably in the neighborhood of Centerville. This channel runs through Mineral Slide and a trifle west of Magalia, up through Lovelock and Sterling, northeasterly through Kimshew Creek and Snow's Mine to what is called Table Mountain. There are apparently two branches of its drainage in the neighborhood of Kimshew Creek, one coming through the Crane Valley. The tributary to this channel above Magalia, known as the Perschbacker Channel, was exceedingly rich and about two million dollars was taken from it. Undoubtedly this channel receives its enrichment from the dyke mentioned before in connection with the Nimshew branch. This dyke runs a trifle east of southerly and may be traced through the country for many miles. The Magalia branch has not been worked out, largely on account of difficulty of operating conditions. There is a great deal of water, and the ground is hard to hold. A very expensive tunnel will be necessary in order to prospect this channel and secure drainage. On the other hand, the Nimshew is accessible by means of short tunnels. Exposures of well-washed gravel are seen along the road from Nimshew to Centerville and breakouts from this channel have been worked at the Oro Fino, Indian Springs, Robbers Roost and Kohl properties, with a reported production of over a million and a half. On Big Butte Creek is a large amount of shore gravel which might possibly pay to hydraulic if the debris question could be taken care of. On the upper portion of the Nimshew branch, near the summit of the West Branch Divide, in Philbrook Valley, and on the Gravel Range are the headwaters of this channel. Tributaries of subangular gravel come in from Carr's diggings and the Westcott Mine.

The Magalia channel can be traced northerly by its rims from west of Magalia through Appleton's and past Doon's Mill, directly under Sterling. North of here, on the southwest side of the west branch is an excellent exposure of the rims and gravel, showing the ancient wash, mixed with subangular gravel of later age which is slightly cemented. This channel continues northeast to Table Mountain.

At the head of the Dry Creek drainage near Cherokee is one of the most baffling channels with regard to origin in the State of California. It is possible that it originally came from the Walker Plains over above Las Plumas to Cherokee and south under Table Mountain. The rolling hills between Oroville and Pentz Ranch are covered with shore gravels and delta gravels from the Cretaceous rivers which debouched along the shores of the ocean. Originally a region of low relief, the uplift of Jurassic times caused very rapid erosion. The Cherokee channel, with a short tributary above Pentz Ranch, is all that is left in channel form of this stream, with the exception of all that buried under the Table Mountain basalt flow and a small section above Las Plumas. Nevertheless, Oregon Gulch, Cherokee Gulch, Morris Ravine, and numerous other gulches have been enriched from the erosion of this channel. The pay channel is over a thousand feet in width and has been hydraulicked at the Spring Valley Mine for a mile and a half in length. Coarse gold was obtained and also some diamonds, which are peculiar to this channel alone. A mile and a half of possible drift ground still remains near Cherokee and probably at least as much under Table Mountain. A narrow pay streak in this channel prospects very well, and a leaner streak about twenty feet above will run about $2 to the yard. Whether drifting can ever be made to pay on this channel is extremely doubtful, due to the cost of handling water and to the fact that the pay streaks are widely distributed. This channel is one of the oldest in the state and in its lower portion may possibly be of preCretaceous age. It extends clear through the South Table Mountain almost down to Oroville.

Dredging ground around Oroville comprises several thousand acres of the present flood plains of Feather River. The width varies from one to two miles. The average depth of the gravel is from 25 to 40 feet. The gravel rests on a false bedrock of volcanic tuff, and is at the present time almost entirely worked out. The upper portion of the Feather River drainage passes from the amphibolites into the Carboniferous slates and limestones through what is known as the upper gold belt of the Sierras. Of the Quaternary gravels of the Feather River little has been said. They are, of course, concentrations of the Tertiary and earlier systems. In the lower reaches of the present north, middle and south forks the Cretaceous and Tertiary channels were evidently almost completely broken down into the present channels and reconcentrated. This accounts for the tremendous enrichment at Bidwell Bar and smaller bars, such as Island Bar, Rich Bar and Big Bar. In every case, the breaking down of an earlier channel is directly responsible for this enrichment. Along the courses of all three forks of the present river are innumerable benches from the level of the river up to five and six hundred feet above it, which would pay to hydraulic in a small way, provided water could be obtained inexpensively. It will, of course, be impossible to enumerate these bars individually for lack of space. The only gravels whose systems will be traced are those of the earlier channels.

The Oroville Basin has been so frequently studied and is so well worked out that little time will be taken in describing it. The reader is referred to the bibliography for the results of many excellent detailed studies of this region.

Commencing in the neighborhood of Brush Creek on the main Pikes Peak highway, a channel, which may or may not be the upper extension of the Cherokee Channel, can be traced through Junction House and Merrimac to the Walker Plains. There seems to be no doubt that this was the master stream which corresponded to the Feather River drainage during Cretaceous and Tertiary times. The wash is mixed, being largely volcanic with a considerable proportion of white quartz, which probably dates from Cretaceous times. This channel for the purpose of convenience we will call the Mt. Ararat Channel. Rising somewhere above Spring Garden it goes across Thompson Creek near the summit of the Nelson Point Road, and through the Clermont Hill. Here it was possibly joined by a short tributary from the east. At Clermont this channel divides, to be reunited at Mt. Ararat. Two different courses are evidently of distinctly different age. One course crosses Bear Creek and comes up on Mt. Ararat to the east of the older channel, which can be traced through the north side of Clermont and out by way of Hungarian Hill, and the head of McFarland's Ravine, where it joins the front channel and breaks out into Willow Creek. The erosion of this channel has enriched Willow Creek from this point down to the middle fork of the Feather River. A tributary comes in across Hartman's Bar where it has been broken down into the middle fork and caused considerable enrichment. An unsuccessful attempt was recently made to mine a hole in the middle fork of the Feather River at this

PHOTO NO. 25. Flume at Morington Mine on Middle Fork of Feather River.

point, and a very considerable amount of money was wasted without determining whether any values existed as a result of the concentration of this channel. This can be traced under Franklin Mountain on the south side of the middle fork. Its course is uncertain beyond Dogwood Peak and may possibly be trending toward Little Grass Valley. The gravel in this channel is well rounded and probably of Cretaceous age with Tertiary wash on top. The front channel has some possibilities for hydraulicking, and the back channel may possibly contain some good drift ground. It is practically virgin.

From Mt. Ararat southwesterly this same channel can be traced by way of the Gravel Range and the Walker Plains above Merrimac. That this was the master stream, draining the present Feather River country in Cretaceous and Eocene times, seems very probable. The gravels of the Walker Plains and on the head of Marble Creek above Merrimac correspond in every particular to those of Clermont Hill. Below Buckeye and Merrimac the erosion of this channel has greatly enriched the gravels of the tributaries on both sides of the ridge into the north and middle forks. Notably this is true of the Mosquito Creek and Sky High Ridges. Below Sky High on the middle north fork there is considerable channel gravel and it extends as far down as Turner's Mill near Bald Rock Canyon.

On top of Clermont Ridge heavy wash gravel, which shows for a distance of about two miles, proves the existence of the above mentioned channel, presumably the equivalent of the present middle fork of the Feather River. A great deal of work was done by an English company several years ago in an effort to prospect this channel. Two long tunnels were driven, both of which in the writer's opinion parallel the channel instead of cross-cutting it and both were too high. No results, either positive or negative were obtained. The proper place to prospect this channel would have been on the north side in the gulches just above the turn-off on the Meadow Valley trail. This is undoubtedly the same channel which runs under Mt. Ararat and can be traced

PHOTO No. 26.   Derrick at Morington Mine on Middle Fork of Feather River.

for over twenty miles. At the head of Gansner Ravine a lower channel goes through which has been prospected slightly by the Laurison Tunnel over the ridge to the east. This channel crosses over the head of Mill Creek, and is said to be traced clear over to Happy Valley. Nothing much is definitely known about it. It is supposed to have come through Hungarian Hill, where it was worked and proved to be fairly rich. It contains very heavy wash.

Tributary to this main channel are two branches which come in from the southeast side. Commencing up near Fowler Peak, and coming down by way of Browns Hill, Sardine Gulch and the ridge just southeast of the present middle fork, crossing near Cascade and Lava Top, and going down as far south as Lumpkin, is a very well-defined channel from which the lava cap has been largely eroded. This channel is probably prevolcanic and of at least Eocene age. There are several miles of possible hydraulic ground upon it, and the whole is covered with excellent timber. The great difficulty is the fact that it will be extremely hard to get water on it under sufficient pressure to operate.

Water would have to be brought from five to ten miles. Dump is excellent. About the only work now being done on this channel is at Spencer Mine above Cascade. This was formerly owned by Mr. Bean. It consists of erosional material from the original channel.

It is entirely distinct from the Franklin Hill and Deadwood Peak channel, which crosses the present middle fork above Hartman's Bar and goes across Willow Creek to the Gravel Range. This same channel has been eroded below Lumpkin and probably enriched the Mooretown, Sucker Creek and Enterprise diggings. A portion of it still remains on top of Kanaka Peak, which has been recently drifted.

At Cascade this channel did not pay, although considerable money was spent in developing it. Higher up at Browns Hill and Sardine Gulch considerable money was taken out.

There is not much doubt that this, like the Mooreville Ridge and the Ararat Channel, formed the ancient Feather River drainage. There is no possibility that either of these channels were drained by the Cretaceous Yuba. Most of this channel could be readily prospected with a drill, as it is on the wagon road as far up as Cascade. A branch of the Western Pacific is now being run to Lumpkin for the benefit of large timber holdings along this channel, which is almost entirely in private hands.

Coming down from Little Grass Valley, a channel follows the Mooreville Ridge in a southwesterly direction. This is probably of at least Eocene age. The gravel is rather light, about fifty feet thick, and in many places is covered with pipe clay and a thin basalt cap. The Dodson Mine is the principal working on this channel and paid very well as a hydraulic mine, although drift operations showed the bedrock pay to be very spotty. This channel is from six to eight miles in length, and appears to be good prospective hydraulic ground. It enriched both the south fork and Lost Creek on either side of the ridge. After reaching the junction of the south fork and Lost Creek, the crossing of this channel is probably through Field's Ranch toward Mooretown. It is not absolutely certain, however, that it did not follow the present south fork channel and that it might possibly have been eroded into it. In the ridge on the southwest side several gulches and creeks, draining into the Feather River, have been enriched with well-worn gold of about the same appearance and grade as the Dodson gold. This is notably the case in Robinson Gulch, which produced over two millions. If this channel goes under Sunset Mountain, it is very deep, as the whole country around Forbestown is slide country. Its presence, however, seems to be indicated by numerous springs which maintain the same level and which flow the year round.

Above Strawberry Valley and southeast of Lost Creek is another channel which was probably drained by the ancient Yuba. It extends from the head of Sly Creek down across Owl and Eagle Gulch and through Clipper Mills. This channel has been worked since the early days and was quite productive. Considerable drift ground remains on it yet. It passes out to the south before reaching Woodleaf and may possibly connect with the fork of the Yuba which comes out east of Challenge.

Coming back to the country around Mountain Meadows, we find a very important channel which has its origin near Prattville. This channel is divided into two parts. They are exposed on the east side of Barker Ravine at Cummings Hill and in the Great Western Power Company's tunnel from Butt Valley to Almanor. Starting near Lake Almanor, the crossing is by way of Prattville and Cummings Hill, where the channel is bent by faulting. From here, around the head of Barker Ravine, the top gravel was hydraulicked fifty years ago and continued around through Dutch Hill. The front channel was all drifted out here, but the back channel remains. Making a sharp, right-angle turn, it passes down under the ridge below Dutch Hill and out through the Sunnyside Mine. The gravel is two hundred feet deep, but only the bottom portion is good pay. It is lava capped all the way down. There is an opportunity for a large drift mine here, working either through the Kelley or the Cameron ground deep enough to bottom the whole channel. At the Sunnyside the pay is quite consistent and from 60 to 220 feet wide. Another channel crosses high up on the head of Chips Creek and Yellow Creek. This is also prevolcanic and goes down from Humbug Valley.

In this same region the Glazier Channel comes down from the present Lake Almanor, on the east side of the north fork of the Feather River, crosses about two miles below the lake, and comes down the west side below Seneca, where it is probably eroded by the present north fork. Lower down under Red Mountain it appears again on the east side of the river. This channel is comparatively recent and starts only sixty feet above the present north fork. Lower down near Seneca it is two hundred feet higher than the present river. In many places, as at the Scott Mine, a great deal of very good hydraulic ground is still left, although as a drift channel, it has never paid well.

Most of the gravel left around Belden, aside from the main Feather River, is confined to Chips Creek, Mosquito Creek and Mill Creek. At the head of Chips Creek is apparently an old channel with very heavy wash. Lower down the creek the channel of the present stream was extensively worked in the early days. The same ancient channel probably crosses the head of Yellow Creek. Mosquito Creek, while very short, has probably more virgin ground suitable for drifting by the small miner than any area in this country. The type of gold now being produced is very heavy and the wash seems to be much lighter than that of Chips Creek.

On Mill Creek are still several large areas of gravel which are unprospected, extending down from near Mountain House to the mouth of the creek. This creek was undoubtedly fed from a channel which can be noted on the Spanish Creek side which passes through Bean Hill and Kanaka Flat. Some of this ground is said to pay very well even yet.

On the main north fork of the Feather and on Indian Creek, which is locally called the east branch, are many small areas of gravel from one hundred to five hundred feet above the river, which would still pay to hydraulic, but are not individually of any great extent. In Round Valley, above Greenville, there is a local concentration of subangular gravel which is said to be rich.

About twelve miles southeast of Quincy, Nelson Creek empties into the middle fork. The bars on either side of the creek have been worked for many years, as well as the bed of the creek. The middle fork at this point was exceptionally rich in the early days and was worked to a great extent, including the higher bars on either side. Apparently there are numerous small areas of gravel still left on Nelson Creek at points fifty to one hundred feet above the bed of the present stream, which are still being mined. Some heavy gold is being taken out on these short sections. Most of these bars can best be worked by drifting. The wash is very heavy. The bedrock varies from a block schist to a soft schist and slate. Practically all of this territory is in the Paleozoic metamorphics. Considerable work has also been done on Willow Creek above Nelson Point, but there is still a good deal of ground which apparently has not been prospected. At Nelson Point the Pauly Ranch undoubtedly contains a short turn of the middle fork which is nearly half a mile in length and which is said to be still intact. If this is the case, there is a possibility for some good drift ground here, as all of

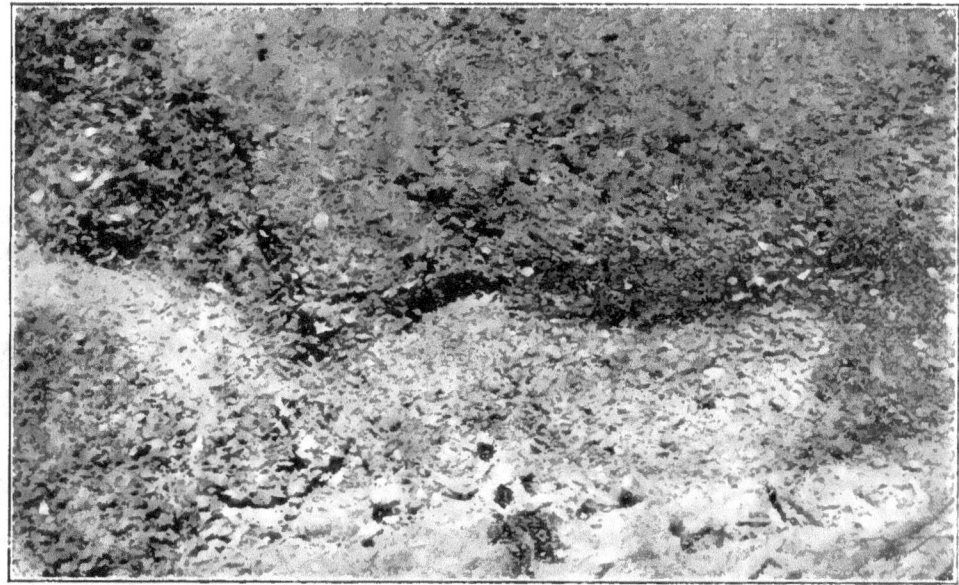

PHOTO No. 22. Gravel Bank at Australia Mine.

the high bars of the Feather River in this region paid very well in the early days.

The region in the neighborhood of Quincy and Meadow Valley has been greatly complicated by Pleistocene faulting, which has made it exceedingly difficult to correlate any of these channels. Meadow Valley, which is believed to have been formed at the close of the Miocene period by a down-throw, was in Pliocene and Pleistocene times probably a glacial lake, fed by streams of Pleistocene age. It is, roughly, an area of four or five square miles, practically all covered by gravel from three to eighteen feet thick. The main feeders of the present valley are Waupause Creek and Spanish Creek. Gravel has been worked on all sides of this valley, and it seems possible that the main valley might have some ground suitable for a very light dredge or for a Ruble grizzly. Water for the latter could be obtained from Spanish Creek for three or four months of the year, in excess of the requirements of the Spanish Creek sawmill, which holds first water rights.

The Australia Mining Company recently operated a mine on Waupause Creek by the hydraulic process. The gravel and overburden were about a hundred feet in depth. With the exception of about six feet of blue gravel on the bottom of the channel, most of the gravel was subangular and was probably late wash. Operations in 1921 gave a total recovery of about two cents per yard. Apparently much fine gold was lost on account of the grade which was given to the sluices. The miners in this region persist in using from twelve to sixteen inches of grade per twelve-foot box, regardless of the width of sluice or quantity of muddy water that may be used. This is all right for heavy gold, but the process used in northern California with five- to nine-inch grades for saving fine gold, according to the writer's experience, produces much better results.

On the opposite side of Waupanse Creek lie the old Gopher Hill diggings, closed down by the Debris Commission about 1895. Most of the debris from this mine was left on the dumps, only the very fine pebbles and silt going down Spanish Creek. This region is very well worked out.

Photo No. 23. Gopher Hill Diggings.

The Australia channel is supposed to come down from the old Pine Leaf diggings, and is about two miles long.

On the southwestern end of Meadow Valley are numerous small deposits of subangular gravel, most of which are probably lacustrine in origin. The most notable of these, Sead Point, is reported to have produced about a hundred thousand dollars. Operations on this end include the Hazel and the Deadwood mines. The Channel Peak Mine on the west side of Spanish Peak may possibly be a continuation of the old channel which crossed at the head of Mill Creek above Belden. This is probably one of the Neocene feeders of the area previous to the down-throw of Meadow Valley. This channel on Spanish Peak is about two miles long, and is covered with a very heavy pipe clay.

On the northwestern side of the valley, coming up Spanish Creek and its tributaries to Mountain House, is a channel system which is prevolcanic, and is probably one of the most important streams of this area. A well-defined channel, notable by the white quartz which predominates, can be traced through from Bean Hill clear over to the north fork of the Feather. It is, of course, broken in many places by erosion. Most of the lava cap has been broken down. A considerable portion of this channel is still undrifted and is the property of the Spanish Peak Lumber Company. Water for operating would have to be brought from Spanish Creek. This channel is about a hundred feet deep, the top portion being white gravel, which at Bean Hill proved to be very poor. Drifting operation in the red gravel close to bedrock appears to show much better values. Following up Spanish Creek below Mountain House there are gulch diggings which were evidently rich enough to be extensively worked, and below Mountain House is Kanaka Creek and Maple Flat, which in 1875 or thereabouts supported many men. Most of this gravel appears to have been a concentration from the main channel which goes over toward the north fork of the Feather. It may have been a concentration of this channel or it may still be possible to develop a good drift mine on Kanaka Flat. The old ditches are still in existence and plenty of timber is available.

It will be noted that, while much of the gravel surrounding the valley is recent and of lacustrine origin, there are still two main channel systems of prevolcanic times which have furnished most of the gold which has been taken from the Meadow Valley and Spanish Ranch district. A reported production of twenty-four million dollars has been taken from this region. On both of these channels areas of undrifted ground are still left, sections of which might be worth while prospecting.

The American Valley, in which Quincy is situated, has been prospected only slightly. The ground is too deep for dredging and too heavy and full of water for drifting except at great expense. Agassiz and Shaw attempted to work the bedrock of Spanish Creek over twenty years ago but met with complete failure after the expenditure of a great deal of money. Shafts sunk by local prospectors in the tributary branch of the valley have given much better results. One shaft, sunk in the northern portion of the valley below Elizabethtown Flats, is reported to have yielded two hundred thousand dollars.

On Rock Creek, which is a tributary of Spanish Creek, and on Slate Creek is the Plumas Imperial property. The main channel comes down between both of these creeks on the ridge. A face 1500 feet wide was worked some twenty-five years ago by Mr. Hazzard, who then owned the property. This face shows that a portion of the channel is a light cemented gravel similar to the Channel Peak, and may have been left there by the major down-throw which caused the formation of Meadow Valley at the close of Neocene times. The remaining portion of the channel is a fine gravel of probably Pleistocene age. The dump into Rock Creek is excellent with good impounding facilities in the basin. There was apparently ample water for operations with a head of 450 feet. This ground is now owned by the Spanish Peak Lumber Company. Above this on Slate Creek two other channels seem to have emptied into the main channel, but were eroded at the point of junction down into the basin, which has never been bottomed. One channel comes down Quigley Ravine and has been worked since the earliest days. The other comes down Slate Creek. In this basin there is a possible chance for a hydraulic mine, as the basin could be bottomed by a four or five hundred foot cut up Slate Creek, and from this basin both Quigley Ravine and Slate Creek gravels could be worked. The basin mentioned formed a natural concentration basin for the gold from these channels. The gravel above this basin, when hydraulicked, is reported to have paid $1.50 a yard. There is also an excellent opportunity for drifting on the Rock Creek side, as the pay in this portion of the channel is confined to a narrow width.

On Mill Creek, east of Quincy, an area of gravel about a quarter of a mile long on the middle fork is now being prospected by a bedrock tunnel some six hundred feet in length from which raises are being made to the ancient channel. Below on this same property is a flat which has never been worked, which looks like a concentration from this same channel. Farther down on the same creek, where it empties into the American Valley, is a large area, which, if the boulders are not too heavy and the ground is not too deep, might possibly have an opportunity for dredging.

North of Quincy, Bushman Gulch, which enters Black Hawk Creek on the southwest, has been worked since the early days. At the head of the gulch the channel extends for over a mile. The whole gulch, including tailings, would probably pay to hydraulic. At the top of the gulch three shafts have been sunk to bedrock at a depth of nearly one hundred feet. This bedrock is higher than the main channel and may be either a fault block or a bench which has not been eroded. In order to tap the main channel with drifting operations, a tunnel 1800 feet long was run about thirty years ago, which is twenty-two feet above the bottom of the channel. A new tunnel at about fifty feet greater depth would bottom this channel and permit drifting virgin ground for nearly a mile which is locally supposed to be good. This tunnel would be rendered much shorter by first stripping the gulch to bedrock by hydraulic methods. Excellent impounding space and dump exist in Black Hawk Creek, and plenty of water is available in the old Bushman ditch. Possibly the best place to get in to the Bushman channel for extended working would be over on the Black Hawk side. A tunnel about a thousand feet in length should give access to the entire channel, as both rims are visible. For the upper ground a short tunnel could be driven below the present shafts. Part of this ground paid from five to ten dollars, but a portion of it is still unworked. Lower down on the channel the gravel is said to be practically intact. This portion would be best reached from the Black Hawk side.

The greater part of the Quincy district is composed of slates and quartzites of the Calaveras formation. Clermont Hill is capped by basalt. West of Clermont Hill the contact is with an amphibolite schist. Farther north the contact is with the Calaveras. All the region from Clermont to the northwest of Thompson Valley, including the rock under the alluvial deposits of American Valley, is composed of Calaveras formation. American Valley and Thompson Valley are covered by recent alluvial deposits. American Valley represents a basin filled with gravel, sand and other sediments, deposited by large creeks emptying into it. The Calaveras formation was deposited in Carboniferous times and is the oldest formation in the district. This is immediately overlain by the youngest formation, which is the alluvial deposit of American Valley.

In American Valley Pleistocene gravels are found between Quincy and Meadow Valley, several hundred feet above the present canyon of

Spanish Creek on the north side of the stream. American Valley was evidently a lake until a short time after the dislocation at the close of the Neocene period. Gravels indicating an outlet are found near Elizabethtown. The gravel corresponds to some small remnants of bench gravel about five hundred feet above the present bottom of Lower Spanish Creek. This outlet was later abandoned by Spanish Creek in the gulch from Elizabethtown to American Valley. Auriferous deposits of later channels are found draining toward American Valley.

Most of the prevolcanic streams of the Sierras appear to have had a general southwesterly drainage into the inland sea of the Sacramento Valley. Upon the present drainage of the Feather River, there is evidence, however, of a stream which flowed northerly and westerly toward Mountain Meadows and was of considerable extent. This has been named the Jura River. Two main branches of this stream have channels which are distinct and well-defined. The westerly branch is first noted in the Mohawk Valley. On the south side of this valley there is evidence of a channel apparently prevolcanic, which is exposed at the Jackson and the Wilson diggings on Sulphur Creek. This channel contains a deep wash without volcanic pebbles, much like the wash of the McCray Ridge but not so heavy. It may possibly be a portion of the same system which was thrown down by the Mohawk Valley fault. More probably, however, it is a portion of the Jura channel, which flows along the north edge of the valley. This channel comes through from above Clio and around by Blairsden. Continuing on under the southern base of Jackson Peak, it extends northwesterly. Up above the lumber mill at Cromberg there is evidence of hydraulicking done seventy years ago on this channel. Below Jackson Peak the channel is well-defined with some intervolcanic wash and some older gravels, extending all along the northeast side of the Mohawk Valley. The channel passes through the Tefft Mine and over the head of Long Valley and Little Long Valley to Grizzly Rock above Squirrel Creek. From there on it makes its way on the south side of Grizzly Valley to the Cascade Mine, and then on down Grizzly Creek toward Genesee Valley. There has not been much drifting on this channel, although the gulches intersected by it have, in almost every case, been ground-sluiced or hydraulicked. A great deal of hydraulicking was done above Lovejoy's at the Cascade Mine. From here on the channel is practically eroded into Little Grizzly Creek which it has enriched all the way down, resulting in the Grizzly Creek placers.

This channel crosses Indian Creek somewhere below Genesee and appears again a mile northeast of Mount Jura at Taylor's diggings, where apparently the southwestern rim has been worked. The gravel here is clearly prevolcanic in character. At Taylor's diggings the channel goes deep under the ridge and crosses the heads of Light's and Cook's canyons southwest of Kettle Rock and through Moonlight Canyon. From here on it empties into Mountain Meadows, where it winds out into a very deep channel which is probably a delta. On the rims of this, and especially above here in Moonlight Canyon and Light's Canyon, much gold was taken out in the early days.

The main branch of the Jura River, however, was a much longer and wider course to the north and east of the one described. This emptied into the ocean much farther north than Mountain Meadows, being about where Duck Lake is now. The lower end is completely covered by a volcanic cap. The head waters of this channel were probably somewhere near Loyalton and are deeply buried under the Pleistocene wash of Sierra Valley. It crossed somewhere above Beckwith and can be picked up above the road on Red Clover Creek a few miles from there. From here on it can be trailed on the southwest side of Red Clover Valley around the east side of Mount Ingells and across Squaw Creek and Indian Creek a considerable distance above Genesee. It passes north of Kettle Rock and through by Diamond Mountain where it has been bent from the Honey Lake fault, thereby giving the impression of a southward flowing channel. It passed west of Susanville and crosses the road not very far from Westwood. Its course can be clearly traced through the Walker Ranch as far as Duck Lake. A branch probably came in at the head of Lone Rock Creek, which crossed high up on Indian Creek. The fact that this is the older course of the river is evidenced by the age of the fossils found west of Susanville, whereas the western branch, described before, shows fossils that are much younger. For this reason, it seems possible that a diversion may have been caused by the Honey Lake faulting, thereby changing the stream from an older to a later bed. Possibly, however, there was a junction of these two channels at the delta near Stockton's Ranch. The immense amount of gravel here is probably more largely due to the eastern branch than to the western, which is comparatively small. This channel can clearly be traced as an entirely separate stream from that of the Mohawk Valley until the possible junction above mentioned.

In addition to the gravels already discussed under the head of the Feather River, there are certain sections of channel which undoubtedly drain southward toward the old Cretaceous Yuba. Commencing above Mohawk Valley on the ridge dividing Jameson Creek and the east fork of Nelson Creek, which are both tributary to the Feather River, there is a channel called the McCray Ridge Channel, which is still very largely virgin, and in many places is heavily lava-capped. There are excellent opportunities for drifting on this channel, and the work at the Sunnyside Mine has proved the values contained therein. The first trace of this channel appears above Squirrel Creek in the Mohawk Valley. Crossing over to the west side of the ridge, it breaks out on the head of Jameson Creek, and can be seen all the way down from the crossing of the Johnsville-La Porte Road for four or five miles down to the Sunnyside Mine. This channel carries a great deal of water and is Cretaceous in its age, the original channel being without volcanic wash. On this ridge there is at least four miles of virgin channel. It is one of the most important channels in the district and, judging from what was done at the Sunnyside Mine above Rattlesnake Peak, it contains good values.

About a mile north of Rattlesnake Peak a flow of the younger basalt crossed the McCray Ridge Channel and forced a large portion of the gravel down the hill. This was discovered in 1882 and worked by hydraulicking as the Sunnyside Mine. In ten years upwards of ninety thousand dollars was taken out in the handling of some twenty-five thousand yards of gravel. There still remains nearly ten times as much gravel here, which seems suitable for hydraulicking. The great difficulty, however, lies in getting sufficient water on the ground to handle the material to the greatest advantage. In the early days whatever water came down in the spring and fall was impounded, and as much as two thousand dollars taken out with twenty days' work for one man. This gravel has undoubtedly been pushed out from the McCray Ridge Channel. A tunnel was run in about 850 feet to tap this channel but was too high. A shaft was sunk at the end of the tunnel ninety feet to bedrock, but there was not a very thick streak of pay gravel. About twelve carloads taken from this streak, amounting to only about four cubic yards, before the water from the channel drove the operators out produced $150. The gold was heavy, as was that in the hydraulic

PHOTO No. 27. Gibraltar Mine, Sierra County.

diggings below, the largest piece being over $3,000 in value, from the Sunnyside Hydraulic Mine. The channel swings southwest across the Burnham ground, adjoining the Sunnyside, and goes on under the east slope of Rattlesnake Peak, as the rims can be traced over half a mile at this point. Beyond the west slope of Rattlesnake Peak is the Gibraltar Mine. It is working on the gravel which is probably another run of this same channel. Considerable heavy gold has been taken out.

On the south fork of Nelson Creek, which drains into the middle fork of the Feather River, there is evidence of a channel which extends northeasterly from Table Rock under Mount Fillmore. In addition to this, the extension of the Hepsidam Channel comes around under Blue-nose Peak and swings easterly toward the McCray Ridge. This has been eroded into Nelson Creek and has gradually enriched it. Near the head of this creek erosion from Bluenose Ridge has produced a large amount of gravel which is suitable for hydraulic mining, and which has been mined in a small way for several years with debris dams. The principal operators are at Bull Gulch, at Red Ravine and at the Standard Mine on the main Nelson Creek. This ground has been worked since the early days and has proved quite rich. At present it is being operated by ground sluices, but a large area of hydraulic ground is still available. The Standard Mine put in a debris dam, but it is now full, and they have discontinued operations.

On Hopkins Creek, Poormans Creek and Dixon Creek there is considerable hydraulic ground, caused by the erosion of the Onion Valley

PHOTO No. 24.   Nelson Creek, Plumas County.

and Sawpit channel, which will be discussed under the head of the Cretaceous Yuba River drainage.

Including all of the available drift mining territory of the possible hydraulic ground embraced in the drainage of the Feather River, it is fairly safe to assume a total of about five hundred million yards, of which probably three hundred million yards is of economic value. The lower portion of the Feather River system drains the area of amphibolite schists and green stone. Higher up, however, we encounter rocks of Carboniferous age in addition to great masses of Tertiary volcanics. The Feather River region, as a whole, has produced many millions of dollars, but at the present time there is very little gravel property in an operating condition. Notwithstanding this fact, there are undoubtedly several opportunities for opening up profitable drift and hydraulic mines. The question of debris storage would probably be solved by building dams on Indian Creek and on the middle fork of the Feather River.

The water rights in this country are excellent, as they drain from the high slopes of the Sierras, on which the snow is densely packed in winter.

REGION OF THE YUBA, BEAR AND AMERICAN RIVERS.

In the consideration of the region around Bangor, although it is drained by Honcut Creek, which is a tributary of the Feather, it is taken under this heading as during the period of its formation it was undoubtedly a portion of the ancient Yuba River delta. A large channel, which is first noted on the Turner Ranch near Bangor, crosses under the town and runs toward Wyandotte, thence turning westerly toward the Feather River. This is locally known as the Blue Lead Channel, and the point at which it joins the Neocene shore gravels is still obscure. It has been extensively mined at Kentucky Ranch, at Bangor and at the Turner Mine, but the main channel is still intact. The gold is rusty and rather flaky and in spots on the upper benches the channel is very rich. The gravel is hard and cemented in the main channel, although some soft ground exists near Bangor. This channel could be drained above Kentucky Ranch, but most of it would have to be worked by a shaft. Much gulch placer work was done near Wyandotte in the early days. Most of it probably came from local seams and pockets which were concentrated in the gulches. The same type of workings appears near Honcut, but apparently most of this ground has been completely worked out.

A good deal of loose, detrital material has been washed for gold at Perry diggings and other places north of Honcut, but most of this material is similar to that around Wyandotte, being probably of Pleistocene age, and deriving its gold from quartz seams in the underlying rock. South of Wyandotte is an exposure of shore gravels that have been washed by hydraulicking, which contain layers of yellow sandstone. In this region there is evidence of shore gravels of two periods which merge into one another.

In the dredging fields around Hammonton are extensive areas of gravel which are undoubtedly a portion of the delta of the ancient Yuba River. These dredging fields have been so often described, and are so near exhaustion, that the reader is referred to the bibliography for details concerning them. These gravels often occur interstratified with rhyolite and andesite mud. For some distance above Hammonton, in the low, rolling hills to the east, there is no further sign of this channel, but about six miles west of Smartsville, and south of the present Yuba River, there is a very good exposure. At Sicard Flat, north of the Yuba River, a large quantity of hydraulic gravel is still available, which has an excellent dump into the present Yuba. Above this, at Timbuctoo, the gravel is all well mined out, but in the neighborhood of Smartsville there are a great many million yards which appear to be excellent hydraulic ground, at Succor Flat. From Smartsville on up to Mooney Flat considerable mining has been done, but millions of yards still remain to be hydraulicked. From Mooney Flat on up to French Corral the major portion of the channel seems to have been eroded, possibly by the present south fork of the Yuba, but west of Rapp's Ranch there is still a little gravel. The more or less continuous deposits from San Juan to French Corral have been hydraulicked almost throughout their entire extent. The main portion of the channel has been worked, but there is considerable gravel on both front and back rims. The maximum depth of this gravel is 250 feet. Although French Corral has been mined since the early days, above it at Sweetland, Birchville, American Diggings, and Sebastopol there is still a good deal of gravel to be mined. At North San Juan is a tremendous body of gravel available for hydraulicking. Most of this ground is tributary to the old Milton Mining Company's ditch. A characteristic of most of the channel from North San Juan down is that the upper portion seems to have been cemented by tremendous flows of volcanic mud.

At North San Juan is a channel which comes in from the northwest on the east side of the north fork of the Yuba River. This extends for several miles on the west side of Moonshine and contains considerable virgin ground, which would probably be worth while drifting, as, wherever the gulches from Moonshine Creek have cut it, they have been greatly enriched. This channel crosses the north fork of the Yuba in the neighborhood of Foster Bar and goes up toward Challenge. West of Greenville some mining has been done on this channel, and a good deal has been done at New York Flat and east of the Clayton Ranch. Whether the New York Flat diggings are connected with this same channel is, however, extremely doubtful, as the gravel appears to be subangular and is possibly of local origin. It may be that, after leaving Greenville, this channel runs almost due north across the lower end of Woodville Creek, as this was considerably enriched, presumably by the erosion of one of the ancient channels. Above North San Juan for about four or five miles the main channel of the ancient Yuba appears to have been practically eroded. At Badger Hill, however, it again assumes enormous proportions, from Badger Hill clear up to North Bloomfield, along what is known as the San Juan Ridge.

This area probably contains the largest possible hydraulic mine yet remaining in the State of California. Gravels are from three hundred to six hundred feet in depth, and bedrock has not been exposed in the center of the channel during practically the entire length of seven miles. This is undoubtedly the largest single body of commercial hydraulic gravel in the State of California. It is mostly owned by one company and contains anywhere from eight hundred million to twelve hundred million yards of gravel, which will run, according to the records of former operations, in the neighborhood of twelve to fifteen cents a yard. In the Malakoff diggings on the upper end of this body of gravel drifting operations in the past two years have shown values running better than $25 a yard on and near bedrock. In 1917 the ground from North Columbia down to Badger Hill was prospected by Keystone drills very extensively.

The values in this ground may roughly be estimated from the fact that the old washings of surface ground near Malakoff from 1870 to 1874 were estimated at about 3,250,000 cubic yards, the yield of which was in the neighborhood of three cents per cubic yard. In 1876 and 1877 about 1,600,000 yards were washed, which yielded about four cents per cubic yard. This, of course, was top gravel. At the same time the company washed about 700,000 yards of bottom gravel, which yielded about thirty-three cents per cubic yard. The bottom gravel extended from bedrock upward about seventy feet. The top gravel averaged from thirty feet to over two hundred feet in depth. The water used in mining this particular section of channel comes from the old Milton ditch and the Eureka ditch, which derived their waters from Big Canyon Creek and the middle Yuba River. At the time that hydraulic mining was suppressed, the total reservoir area of this company contained a capacity of about 2,200,000,000 cubic feet. The length of the ditches of this company was 157 miles. The length of the ditch of the Milton Company, which operated below, above French Corral, was about 80 miles.

The tailings from this tremendous body of gravel were mostly discharged in tributaries of the south fork of the Yuba and the middle fork. By the building of a dam just below the junction of Deer Creek and the middle fork, and the construction of a similar one below the junction of Shady Creek and the south fork, practically all of the tailings of this area could be impounded for the next hundred years. The channel is probably at its widest near North Columbia. Gravel at this point is over four hundred feet in depth. A junction of two important streams occurs at a place near North Columbia. The total area of the gravels between Badger Hill and North Bloomfield covers about eight square miles. Gravels in the deepest trough, which is exposed at Badger Hill and at Grizzly Hill, are very coarse and made up largely of metamorphic rocks, while the deep gravels which spread out over the benches are largely composed of white quartz and are much finer. Near the surface, but at the base of the lava cap, is a great deal of sand and light-colored clay. In the neighborhood of North Columbia a great deal of surface work has been done and about 150 feet of gravel has been washed off. The deepest part of the deposit at Grizzly Hill will require a considerable expenditure for bedrock tunnels before working. To quote from the report of the U. S. Geological Survey on this area:

"At North Bloomfield the exposures are mostly in the hydraulic bank along the center of the channel. The bedrock rises north and south of the main channel. Across the bottom it is nearly level for three or four hundred feet. The deepest blue gravel is 130 feet thick. This is capped by heavy bodies of light-colored clay and sand, interstratified with fine gravel, and near the top occasionally also with andesitic tuff. The clay and sand may reach 150 feet in thickness. This is again covered by six hundred feet of tuffaceous breccia. The lower surface of the breccia is uneven, as shown by the fact that sand and clay crop out a short distance east of the Dorbec Mine. About one mile north of North Bloomfield the channel forks again below the lava. The main fork has its inlet from the lava ridge north of Backbone House, where the configuration shows the existence of a deep channel, along the center of which Bloody Run has excavated its canyon. Gravels, capped by heavy masses of slide clay, are here exposed.

"Hydraulic mining has been carried on at North Bloomfield on a very large scale. The excavations extend for 5000 feet and are five to six hundred feet in width. The banks are as much as five hundred feet in height. The deposit has been opened by a bedrock tunnel 7874 feet long, starting from Humbug Canyon. The sum of $3,000,000 is said to have been expended upon this tunnel, the water supply and other preliminary work. Shortly after the completion of the tunnel hydraulic mining was suspended by injunction of the courts and since then the only gravels worked have been those the tailings of which could be impounded before reaching the river."

The evidence upon the channel extending from North Bloomfield down as far as French Corral shows very plainly the importance of glacial action with regard to the distribution of gravels and pay. The theory of pre-Tertiary glaciation has not been generally admitted, but it seems hard to explain the distribution of the gravels in this particular area without admitting the probability of glacial action on these channels. The Cretaceous and Eocene periods extended over a period variously estimated at from one million to five million years, and during that time, previous to the period of extensive vulcanism, there must have been many climatic changes, caused by the shifting of ocean currents and pressure areas which would have given the Sierras a climate similar to that prevailing at present in Alaska and British Columbia, entirely independent of any general shift of the Polar ice cap, such as occurred in Pleistocene times.

The evidence and positive proof of local glaciation in this channel is so pronounced that it can not be overlooked. Discrepancies in bedrock elevations in the down-stream course of the channel may be in part, although not altogether, explained by the differences in datum from which these elevations were obtained; but the presence of bedrock

Photo No. 32.   Characteristic Glacier on Alaskan Coast.

fluting and of single and double U-shaped channels; of heaps of heavy boulders of varied character which could not have been carried so far from their source by water alone; and of the immense white quartz channels in which all the metamorphic wall-rock has been ground up and transported out by water; in which the grinding would have been impractical without the aid of ice action; and the fact that the depth and width of these channels was often increased irregularly; all of these phenomena indicate the deposition of much of this material and the hollowing out of the channels by local glaciation. At the present day similar phenomena have been observed by the writer in Alaska, especially in the streams of water discolored by blue mud, which come out from under the major glaciers in the Cathedral Mountains.

That water sorting followed and was intermittent with this action is also indubitable, as the shingling and sorting of the gravel shows beyond doubt that these were the courses of ancient streams. One thing may well be noticed; that is the relative elevations of the lava and gravel contacts. It may well be possible that a series of readings taken on these would show the original grades of these streams far better than bedrock elevations would, even when properly correlated.

Somewhere between North San Juan and Badger Hill the main channel is joined by a tributary coming down from the north. No gravel is noticed in place on this tributary until Camptonville is reached. Here a wide, deep channel has been thoroughly worked out. Above this, at Galena Hill, at Youngs Hill and east of Oak Valley, considerable gravel is still left. The wash is mostly white quartz and not particularly heavy. At Jouberts and at Indian Hill a large amount of gravel is still available for hydraulicking, although the central channel at the former place has been pretty well cleaned out. At Brandy City and at Council Hill are still large quantities of gravel, but the flows of lava mud, lying closely on top, have in many places encroached so heavily on the gravel that it does not pay to work it.

PHOTO NO. 31. Gravel at Scales, Sierra County.

Going northerly from here through Union Hill, which is almost entirely hydraulicked out, the channel extends to Scales.

The Brandy City Mine, which is now being worked, still contains about half a mile of unworked ground along the channel. There is a very heavy top of volcanic mud and from fifty to one hundred fifty feet of fairly good gravel. There is a good water right on Canyon Creek amounting to about three thousand inches. At Indian Hill on the other side of the north fork an interesting development of cemented gravel is seen. Ferric-sulphate is still forming and cementing the gravel at the present time. At Scales, below the junction of the Port Wine-La Porte branches there is very heavy wash, which is largely inter-volcanic. This underlies the original white gravel which is mixed with fine andesitic pebbles toward the surface. There is a tremendous area of unworked gravel here. Mining is still being done in a small way, both hydraulicking and drifting. From Scales on up the Port Wine Ridge, the channel is almost intact and is largely virgin. At Mount Pleasant, above Scales, is the junction of the La Porte and the Port

Wine channels. There are from sixty to eighty million yards of gravel here which would make one of the easiest hydraulic deposits in this state to work. Mount Pleasant and Poverty Hill may be considered as one property, although the Poverty Hill portion is supposed to be considerably better gravel. This is one of the most important hydraulic deposits in the state. About two hundred feet in depth the bank is composed of small white quartz gravel with an unknown width, which is presumably, at least, half a mile. The channel extends for nearly two miles. The center of it has been worked for about half a mile, but the balance is practically virgin. It is eminently suited to hydraulicking, as the values run clear up to the top soil and the bedrock pay, judging from the history of the early mining, is undoubtedly good. The Poverty Hill branch of this channel extends on up through Secret diggings to La Porte. Both of these places are pretty well worked out, and were among the richest gravel deposits ever known in the State of California. Above La Porte the main channel is practically intact, although some prospecting has been done on it at the Bellevue Mine. As the channel at this point was continuously crossing a belt containing quartz seams and stringers which were rich in gold, the values appear to run almost uniformly through the channel. The Bellevue Mine itself probably contains from four to five miles of virgin channel, which should be among the best drift ground yet remaining in the State of California. Passing under the Gibsonville Ridge, where a great deal of hydraulic ground was developed in the early days, it continues on up to Hepsidam and Whiskey diggings. It is almost altogether composed of white quartz gravel wherever it is exposed, and the bottom strata have almost invariably made excellent drifting ground wherever it has been tapped. From Hepsidam the channel continues northwest and east of Pilot Peak. It turns sharply to the east, passes through around to the north of Bluenose Mine and swings over on the west side of Nelson Creek, the tributaries of which have nearly all been enriched. From here on this channel has already been described under the Feather River drainage. Another branch of this channel comes down from the upper end of Little Grass Valley and enters it somewhere in the Bellevue ground, not far from the Thistle shaft. The upper end of this channel is now being worked at the McFarland Mine and is paying very well. The gold appears to be distributed through the bottom ten or fifteen feet of the gravel with a great deal of uniformity. Still higher up, around the eastern slope of Pilot Peak, we have the Onion Valley channel. This is one of the richest feeders of the Gibsonville-La Porte channel that is first noted above Washington Creek near Golden Gate. From here it crosses through the old Sawpit diggings, where a portion of the channel was faulted down for about a mile and was hydraulicked at Richmond Hill. Richmond Hill was one of the richest hydraulic mines in the state at the time of its operation. The whole flat above Sawpit has been drifted out, although the channel was pretty well spread over a large section of country by the faulting. Passing around the head of Weddon Ravine and directly under the main ridge between Dixon Creek and Onion Valley Creek this channel is seen again at Red Slide Hill, where it swings around to the south and joins the main Gibsonville Ridge channel. This channel has been eroded into Dixon Creek and has enriched it greatly. The channel is

very near the top of the old Cretaceous divide and is evidently of
Cretaceous age. Between the old Australia tunnel and the entry above
Sawpit Flat there is probably a mile of virgin ground which should be
some of the best drifting ground left in this state, judging by the
records of the old Sawpit and Richmond Hill mines. On the Onion
Valley side two tunnels, the Pioneer and the Weldon Ravine, have
been run to tap the channel, but to date they have only encountered
the bench gravels in raises. On the Dixon Creek side, the Australia
tunnel has been driven in for over 1500 feet with raises which have
also tapped the bench gravel. Undoubtedly between these three tunnels
is a deep channel which should pay to work. Onion Valley itself, aside
from the main channel, appears to have been a recent glacial valley,
but a large part of the detritus in it is eroded stream gravel from the
channel above mentioned.

On the south fork of the Feather River in Little Grass Valley is an
area which appears to have some possibilities as dredging ground. A
branch of the Gibsonville channel, as stated above, runs across the head
of it and may possibly have enriched it. The wash is not very large,
but there is too much water for drifting operations without driving
a long drain tunnel. This side of the ridge has not been very extensively mined, although some placer work was done lower down in the
early days. Apparently three channels run transversely across the
valley. The most easterly is the north branch of the Gibsonville-
La Porte channel mentioned above, with its west rim near Kinsey
Ravine. Farther west is another channel several hundred feet wide
also running northerly and southerly on which the Tombs incline was
sunk in early days. Both rims are well-defined and from the strike of
the channel it appears to join the main Gibsonville Channel at Dutch
diggings above La Porte. This gravel has a slight proportion of intervolcanic pebbles.

Still farther west is a very large channel which runs nearly parallel
with the direction of the valley. This has been slightly prospected by
Keystone drills. It contains a large proportion of volcanic material
and runs out considerably to the west of the other channels. It is quite
possible that the drainage of this channel is down the Mooreville Ridge
through the Dodson Mine.

The south fork of the Feather River runs directly across all these
channels, and could be used to supply water for hydraulicking if a deep
bedrock cut were run from the outlet of the valley. The gravel remaining under the first layer of pipe clay, which is about one hundred to
one hundred fifty feet deep, might then be dredged and the whole thing
later converted into a reservoir or power site or else, by making the cut
deep enough to drain the valley, the whole thing might be hydraulicked.

At Gibsonville a large area of ground is still available for
hydraulicking. The same channel passes on under the Gibsonville
Ridge and comes out at Bald Mountain. At Dutch diggings and on
Bald Mountain considerable gravel is still available for hydraulicking.

Coming back to the junction of the Port Wine Ridge Channel at
Mount Pleasant, it can be traced to the northeast under a heavy andesite cap for a matter of about ten or twelve miles. At the old Iowa
shaft north of Mount Pleasant good gravel was found in the early days
but was not prospected far on account of water. The area comprising

the Poverty Hill, Mount Pleasant and Scales deposits probably contains about 150 million yards of gravel. The Slate Creek water could be brought down the Port Wine Ridge and used to good effect in mining this gravel. It would undoubtedly make one of the largest and best paying hydraulic mines in the state, although fitting up would be expensive on account of dump and drainage problems. Undoubtedly a long drain tunnel would have to be run from the Slate Creek side to the lowest point in the channels which join on these properties.

PHOTO NO. 28.   Basalt intrusions at Port Wine,
Sierra County.

Although the channel is more or less intact from the Iowa shaft almost up to Port Wine, it is heavily lava capped, and would probably have to be drifted. At Port Wine there is an intrusion of the younger basalt which is very interesting. From Port Wine up to Grass Flat the channel is more or less exposed. At Gardner Point is considerable hydraulic gravel, and at St. Louis is a very large body of gravel which is apparently exceedingly low-grade, the best of it having been worked out by the early day miners. From here on the channel passes up through Howland Flat, where it was extensively drifted, and swings around under the andesite at Potosi, breaking out again at Poker Flat where considerable work was done on it. There is a possibility that the Poker Flat diggings are not connected with Potosi, but that the drainage was down the other way through Deadwood toward Morristown. Almost parallelling the channel, which extends from Potosi to Grass Flat, is another channel which is buried deep under the lava at Mount Fillmore and Table Rock. This channel has been drifted in the neighborhood of Table Rock but is largely virgin from here on down. As it is much narrower than the Howland Flat channel on the west, it is quite possible that it may be considerably better drifting ground. This channel

PHOTO No. 29.   Poverty Hill, Sierra County.

appears at the Pioneer Mine near Grass Flat, but instead of joining the front channel and going down toward Port Wine, it swings out to the south through Happy Hollow and Morristown, where it is joined by a tributary coming down from Deadwood. From Morristown it goes through Craig's Flat to Eureka, where it will be discussed later under the head of White Bear Channel.

PHOTO No. 30.   Close-up of Poverty Hill Gravels.

The headwaters of the White Bear channel are found somewhere in the neighborhood of Deadwood Peak. From here it runs southerly under Saddleback Mountain and Firtop, passing through the Cooper Ranch under Monte Cristo. It was extensively drifted at the old White Bear Mine above Monte Cristo and proved exceedingly rich. The characteristic of the gravel of this channel is the large amount of white quartz boulders. From here the White Bear channel runs southwesterly and is eroded into Goodyears Creek clear down to Goodyears Bar where it caused an extensive enrichment. The upper portion of Goodyears Creek and also part of Eureka Creek were greatly enriched by a branch of this channel coming down from Happy Hollow through Morristown and Eureka as above stated. The junction was probably on Goodyears Creek, a mile or two above Goodyears Bar. Below Goodyears Bar this channel was joined by a tributary coming down from McMahons through the St. Charles Ranch. From Goodyears Bar it goes on southwesterly to the east of Snowden Hill and Humbug Creek. On the head of Little Humbug Creek the lava cap was sufficiently removed to allow considerable hydraulicking. This channel crosses Oregon Creek about two miles south of Mountain House and goes through the ridge to Kanaka Creek. On the south side of Kanaka Creek it continues through the ridge, crosses the middle fork of the Yuba River and swings westward toward Grizzly Ridge. Here it turns southerly to enter the main Columbia channel somewhere east of Tollhouse.

The channel mentioned under the Feather River system as the McCray Ridge channel comes down under the east side of Rattlesnake Creek and runs southerly on the east side of Rattlesnake Creek and on the north fork of the North Fork of the Yuba River. Coming down the ridge between this fork and the middle fork it swings westerly to Monte Cristo where it crosses the White Bear channel. At Monte Cristo a great deal of drifting was done and the channel proved exceedingly rich. From here the channel flows almost straight south under the ridge west of Downieville until it crosses the north fork about two miles east of Goodyears Bar. From here it is eroded into the Rock Creek drainage, but appears again at the Kirkpatrick Mine near the old Henness Pass road. A great deal of drifting has been done upon the channel at this point, and a large amount of low-grade gravel uncovered. From here on it passes through Forest and proceeds southward parallel to the Alleghany channel, thence swinging southwesterly toward Grizzly Ridge.

Another run of this same channel, starting on the ridge between the north and middle forks of the North Fork, comes down through the old Craycroft Mine and is eroded directly into the Yuba River at Downieville. The concentration of this channel into the bars around Downieville is responsible for the millions of dollars which were taken from this place. Crossing southerly through Slug Canyon, it is found again at the Ruby Mine and the old Bald Mountain, where it was extensively drifted and much rich gravel taken out. Here it joins the other channel a short way above Forest City. A short tributary coming down from the east joins it at this point also. Still farther east of this on the head of Secret Canyon another channel comes down through the Pliocene Mine, crosses through the ridge on the head of Little Kanaka Creek and joins the Forest channel at Alleghany.

From Alleghany down this channel was extensively worked clear through Chip's Flat to Minnesota and proved exceedingly rich. Crossing the middle fork to Orleans from this point on down it was extensively hydraulicked to Moore's Flat. Entering the ridge again above Moore's Flat it crosses the head of Bloody Run, and comes through west of Backbone House into the old Malakoff diggings where it is now being drifted by a local company. From the old Derbec Mine immediately above North Bloomfield a channel can be traced through the northern edge of Relief Hill up by Snow Point where it crosses the middle fork of the Yuba River. From here on up to American Hill there is still some virgin drift ground. Two tributaries enter this channel on the east and west sides of the camp of American Hill. From American Hill toward the northeast this channel can be traced through the lava capped ridge on the north side of the middle fork clear up to Milton. At Milton a fair sized body of hydraulic ground was formed from the erosion of this channel. From Milton it can be traced northerly to the Hilda Mine southwest of Milton Creek, and it is again picked up on the opposite side of the north fork at the Thousand and One Mine. A branch coming in from the Deer Creek drainage meets it at this point. The main channel goes along the ridge west of Williams Creek, crossing the north fork of the North Fork below Bassett's. Erosion from this channel on Williams Creek has created a very promising body of hydraulic gravel. Following up Sardine Creek, where it is mostly eroded, this channel is again picked up below Gold Lake. At this point the channel crosses a pocket belt which has greatly enriched it. Above here it is joined by a tributary running westerly from the Haskell Peak region. The main channel crosses through Church Meadows, northwesterly by Gold Lake toward Mt. Elwell. From here its course is rather uncertain, as it was undoubtedly affected, so far as elevations are concerned, by the dropping down of the Mohawk Valley. Judging by the enrichment on Jamison Creek and by the exposures above the Little Jamison Mine, it is probable that its course up stream was in a northerly direction for a matter of seven or eight miles.

On this channel much virgin ground remains to be drifted all the way from Mt. Elwell down to American Hill. It is possible that some of the gravel may be worth while drifting, but the pay is exceedingly irregular and very spotty.

A northward flowing tributary on the east side of English Mountain and Findley Peak joins this channel near Milton but is of no economic importance.

Below Snow Point this channel is joined by a branch which comes down from Pinoli Peak through the granite belt north of Graniteville. This is joined by another tributary from Bowman Lake. Gravel in this channel is practically all granitic wash and will probably not pay to drift. This channel can be traced through Shands down to Relief Hill, where it is joined by another minor tributary which parallels it on the south. None of these latter channels are of economic importance.

From North Bloomfield east a very important channel passes through Relief, where it has been extensively hydraulicked and considerable hydraulic gravel remains. Following southeasterly from Relief, the channel has been eroded by the south fork of the Yuba but can be picked up again at Alpha. From here on its course through Omega can be followed by immense bodies of hydraulic gravel. From Omega it follows the head of Diamond Creek and crosses the south fork of the

Yuba again at Langs, where it can be traced up to Zion Hill. This channel is joined by another tributary coming through Lake Valley north of Emigrant Gap.

At North Columbia this main system of channels is joined by an equally large tributary which comes from the south by way of Grizzly Canyon. This channel, in the extent of its drainage, is probably quite as large and quite as important as any of the northern ones. Following up the stream in a southerly direction tremendous masses of low-grade hydraulic gravel are encountered near Blue Tent and on the Rock Creek drainage. Apparently one branch of this channel passed through Scott's Flat to Quaker Hill. There is an enormous mass of low-grade gravel at Scott's Flat. The other branch of the channel passed through Galbraith and crossed above the north and south forks of Deer Creek. Below this they were joined by an unimportant southward flowing tributary. Joining the course of the old channel again at Quaker Hill enormous masses of hydraulic gravel are again developed. Presumably, in the area from Scott's Flat to Gold Run there is in the neighborhood of a billion yards of gravel. At Hunt's Hill, Red Dog and You Bet extensive mining operations were carried on in the early days, but a tremendous amount of gravel still remains to be developed. Coming up from Tollhouse above Colfax a tributary, which has mostly eroded into the Bear River, joins this channel near Little York. The main channel, however, swung easterly to Dutch Flat and southerly through Gold Run.

In connection with the effect of this channel upon the enrichment of Bear River, it may be stated that the hydraulic operations of Greenhorn Creek and the main Bear River have produced a tremendous mass of tailings which have been variously estimated to still contain from 20 to 40 cents a yard in value. Several years ago a report on this subject was made by two or three of the most prominent engineers of the time, in which recommendations were made for a tunnel which was to be driven through from Bear River below the junction of Greenhorn Creek to Secret Canyon, a tributary of the American.

The idea of this project was to permit the working of the enormous accumulation of tailings on Bear River and, on account of the difference in elevation between the bed of Bear River and that of the north fork of the American, to thus create a dump which would permit of the disposal of the tailings in their reworking.. This project was afterward abandoned, doubtless because it could not be satisfactorily proved whether the value in the tailings were sufficiently great to permit of a profit in the operation. A map showing the location of the principal properties in the neighborhood of Dutch Flat, Red Dog, Little York and You Bet is herewith included. This map also shows the location of this proposed tunnel for reworking Bear River tailings.

The lower reaches of Bear River, especially in the rolling land above Wheatland, have resulted in the concentration of a certain amount of delta gravel, which was worked in the early days for many years with satisfactory results. Some unsuccessful dredging has been done on the lower reaches. Owing to the lack of water, most of this work was done by means of rockers and short sluices immediately after and during the rainy season. For many years some of the old-time prospectors have been working the short turns and bars of the Lower Bear River for the fine gold which is annually brought down from the erosion of the south fork of the Neocene Yuba in the neighborhood of Dutch Flat and Greenhorn Creek. This area undoubtedly contains at least a billion yards of economically workable gravel.

At Dutch Flat this channel is joined by a tributary from Remington Hill through Lowell Hill, where it is joined by a minor tributary; thence through Liberty Hill and the Polar Star to Dutch Flat. There is still a tremendous amount of good hydraulic gravel, ranging from 10 to 25 cents a yard, on this channel. At Dutch Flat this channel is also joined by what is known as the Alta channel. This can be traced along the line of the Southern Pacific Railway clear to Blue Canyon, where there is a very considerable body of hydraulic gravel. Another branch comes down from Texas Hill through Shady Run and joins this channel. Still another minor branch comes across the north fork of the American River through Euchre Bar and joins this channel at Alta. Following the course of the main channel from Gold Run southerly through Wisconsin and Indiana Hill and, crossing the north fork of the American River, it appears again in the form of immense masses of hydraulic gravel at Iowa Hill. It is here joined by a tributary from Succor Flat through Monona Flat and Roach Hill. At Iowa Hill, extending southerly through the branches of Shirttail Canyon, are still tremendous bodies of hydraulic gravel. From here on the channel can be traced southerly through Yankee Jim, where there is still considerable hydraulic gravel and where drifting operations are now carried on. From Forest Hill on, this channel presents a most baffling situation, due to the tilting movements at the close of the Eocene period, which gradually changed the direction of flow.

The original system, working upward from Yankee Jim, came through the Dardenelles and Forest Hill by way of Bath to Michigan Bluff and up through Turkey Hill and the Hidden Treasure to Damascus and Red Point; thence extending around by Westville to Secret Canyon.

Another branch, which has been greatly eroded by the American and Rubicon rivers, runs northeast to Devils Peak and across through the Ralston Divide.

The lower reaches of the middle American have been successfully dredged, but are now worked out. The enrichment undoubtedly came from the erosion of this system.

At the close of the Eocene period, with the beginning of the heavy rhyolite flows, this distribution was materially changed. Instead of going north by way of Yankee Jim's the channel turns southwest down the present middle fork, where it has been greatly eroded. An intervolcanic channel appears on the Ralston Divide and numerous parallel ones below it, all converging towards Todds Valley and Peckham Hill.

A careful study of this region, made by the writer during 1922, was published in the October issue of 'Mining in California,' and will be quoted bodily herewith. In many respects the writer has found reason to differ with the channel system as quoted in the Colfax folio of the U. S. Geological Survey. In the main, however, the work of the Survey has been so thorough and so painstaking that it seems of little use to revamp any of the information contained therein. For this reason it seems best to quote also from the report upon the auriferous gravels given in the Colfax folio. Following is the quotation from 'Mining in California:'

"From both the historical and the economic standpoint, the buried channels of what may be termed the Cretaceous-Eocene south fork of the ancient river system which drained the major portion of the present American and Yuba rivers territory, are among the most important in the state.

"From Yankee Jim on northwesterly, through Iowa Hill and Dutch Flat to its junction with the north fork at Columbia, the course of this channel is well defined, and admits of little argument. From Yankee Jim east, however, there is a vast difference of opinion with regard to the correlation of this stream and its tributaries.

"This is largely due to the change in drainage of the system which occurred toward the close of the Eocene period and contemporaneously with the earlier rhyolite flows. It seems undoubtedly to be the case that a large block of the western slope of the Sierras was tilted and dropped at this time, and comparatively abruptly; because a drainage was established toward the southwest, and approximately down the course of the present middle fork of the American, along lines which it retains to this day; although a gradual subsequent uplift during Quaternary time has altered the course of present day streams, and heavy erosion has taken place. Coincident with, and immediately following upon, this shift, new channels were cut. The flows of lava pouring down the western slope at intervals, in part following the old stream beds and in part obliterating them and transporting their contents to other courses; the dams of volcanic ash temporarily created and later released like tremendous 'self-shooters,' with all the volume of pent-up torrents behind them to cut new courses; and the comparatively quiescent periods between lava flows when the new streams pursued, perhaps for thousands of years at a time, new and wandering courses, distributing both new and ancient wash according as they were cutting lava or ancient stream beds: all of these influences combine to produce perhaps a unique condition, and certainly one of the most puzzling and complex situations with which the gravel miner has ever had to contend.

"In working out this system, it is absolutely necessary to make one sharp distinction primarily; that between the tributaries of the old Cretaceous-Eocene, northward flowing stream, and those of the post-Eocene, southwesterly course.

"In doing this, it is impossible to correlate exactly. Drainage lines as followed out may not have been contemporaneous by a matter of a thousand years, or even several thousands; but along broad general lines, it seems perfectly clear that the system is well established.

"To begin with the prevolcanic system. Following upstream from Yankee Jim through Forest Hill, Mayflower and Bath to Michigan Bluff, we have an important tributary coming down from the north which we may follow up through Gas Hill, Turkey Hill, Hidden Treasure and Sunny South to Damascus and Humbug Bar, with a tributary coming in by way of Red Point which may be traced below Westville clear up to the head of Secret Canyon. This evidently had a divided course at a later period, crossing over through Black Canyon and being joined by a tributary from Whiskey Hill, and later resuming its course near Westville. Another tributary to the Hidden Treasure branch came in by way of Deadwood. This in turn may be traced through Last Chance, American Hill and north of Antone Canyon up to Canada Hill and Sailor Ravine.

"Coming back to the junction at Bath, we have the old Breece and Wheeler channel. For some distance this has been eroded; but in general it followed the middle fork to the Rubicon and up Long Canyon nearly to Wallace, where it crossed to the south side and goes through what is now Nevada Point to Little Grizzly; thence up Pigeon Roost to the eastern slope of Devils Peak, where it is very deeply buried under subsequent lava; thence through and across the upper portion of Long and Wallace canyons to the northeast side of Jerry's Canyon, where it went through the Ralston Divide and its entrance may be seen above the Little Crater. From Jerry's Canyon on down another run of this same channel may be traced around Blacksmith Flat by way of the old Clydesdale over to the Nevada Point side, where it joins the old course. Following up from Little Crater, it crosses the middle fork to Duncan Creek, where it is joined by two tributaries: the Blue Eyes, coming down from Deep Canyon; and the Glen, coming down from Flat Ravine. Continuing north through the Hard Climb to Chalk Bluffs, it goes up the middle fork toward Castle Peak.

"Again coming back to a junction southwest of Bath, we have another major fork. Crossing in a southerly direction near Volcanoville, we follow it upstream southeasterly toward Twelve Mile House and along the Georgetown Divide through Fornis Ranch, across Pilot Creek to Eleven Pines, and on the south side of the Rubicon through Uncle Tom's and thence easterly and northerly by way of Mount Mildred to Tahoe. A tributary comes from the southwest near Robb's Peak and joins this near the eastern edge of the Placerville quadrangle.

"South of this area, we have tributaries running into another great stream, which was at this time the equivalent of the present south fork of the American, with regard to territory drained. Commencing near the head of Rock Creek and also the head of Silver Creek we have two channels which drained southerly, and are almost eroded except for fragmentary patches until we get to White Rock Canyon and to Badger Hill, respectively. Their relation to the intervolcanic streams will be shown later.

"Such was the general situation during Cretaceous and Eocene times. These channels had planed down the metamorphic rocks of the Sierras for probably a million years or more. In their beds was the source of most of the present day enrichment of the placers of this portion of the Sierras. Like gigantic sluice boxes they stretched their shining lengths over the rolling foothills of the Sierras, wending their way toward the northward flowing tributary of the Cretaceous Yuba, and contributing to the enormous masses of gravel already deposited in the glacier-hewn channels of what is now Quaker Hill and Blue Tent. Whenever these streams crossed a rich pocket and seam area, like that running northerly through Georgetown and Michigan Bluff, they were enormously enriched; but the heavy gold gained from these areas did not travel far, and hence these channels were, like those of the present day, extremely spotty. The fine gold was far more generally distributed.

"The beginning of the era of intensive vulcanism is generally placed at the latter end of the Eocene period. The first manifestations of it were extensive showers of volcanic ash and pumice, which coated the flanks of the Sierras for many feet in depth. Probably the source of these eruptions can be traced to two or three vents; and from the present topography it is rather hard to estimate their positions. These vents covered the Cretaceous summit of the Sierras and its flanks for over a hundred miles in length with a mass of steaming mud and lava.

"As the thick coating of ash was gathered up by the tributaries of the region, it collected in enormous masses along the main channels, causing them to dam up and change their courses. Then came the first flows of rhyolite, and a period of comparative quiescence for a few thousand years. New cement channels were formed, cutting the old ones and changing the drainage lines; but in general the course of the drainage was the same.

"Next came the enormous eruptions of rhyolite and other acid lavas, together with the older basalts on the north. Pouring down the mountain sides in slow-moving masses, converting whole rivers into steam and causing tremendous explosions along their flanks and beneath the viscous masses, they cooled and solidified, raising the enormous rock masses that we now know as the Ralston Divide, the Nevada Point, and the Georgetown Divide, which were greatly increased in their relief by the later erosion of Quaternary streams. Coincident with this came the dropping down of the area south and west of the present middle fork of the American. A new drainage to the ocean was established in a relatively short time; the then shore line being at the edge of the Sierras, and the diversion point being somewhere near the Dardanelles.

"Intermittently, then, was created a whole new system of Tertiary streams, which may well be termed the sluice robbers of that and even of a later period.

"The effects of this sudden change are well shown at Todds Valley. Enormous masses of rhyolite lava, alternating with flows of steaming, viscid mud, tore down the old courses of the white quartz and metamorphic channels and picked up their contents bodily until, losing their grade and their momentum, they deposited them in this flat space. Quaternary erosion did the rest, the gold concentrated, and the result was the enrichment of Todds Valley. The gold originally lay in the channels of the old Cretaceous river; but these sluice robbers picked it up and redeposited it. Another similar drainage was established by way of Peckham Hill. Passing east through Volcanoville, we can follow another stream of this system which is joined by a tributary coming down the Ralston Divide. This latter picked up its enrichment from the crossing of the old channel near and below the Goggin diggings. Gathering some legitimate enrichment in crossing the pocket belt, but for the most part containing what gold it had taken from the older channel, this meets the Volcanoville Tertiary, which heads somewhere east of Bacchis. Together they form a larger channel which crosses Otter Creek to Gravel Hill and Jones Hill, and on into the present middle fork, where they were in turn eroded. At Jones Hill, or near there, this main channel is joined by another Tertiary stream which comes down from Seven Mile House through Bottle Hill. This also has another course through Three Mile House and southwest of the Georgia Slide, where it doubtless picked up some legitimate, primary enrichment. The Ralston Divide branch also had another tributary which later buried Nevada Point under a tremendous lava flow.

"To the southward, via Kentucky Flat, Tiedemenn's, and Tipton Hill, still further flows of volcanic mud and ash, as well as rhyolitic lava, poured down the Rock Creek drainage toward Placerville. Later, this channel, both in its Cretaceous and Tertiary form, was eroded by Rock Creek between Tipton Hill and White Rock Canyon. A few traces of the gravel remain on widely separated points. The same condition was duplicated both as to formation and erosion by Silver Creek in the channel which came down from Eleven Pines by way of Mundy's and McManus to Badger Hill.

"While these Tertiary channels may have accumulated a little gold for themselves, in the main they secured what wealth they have from their crossing and intermingling with the older Cretaceous drainage already described. In the main, they do not pay to drift, except near the sources of their enrichment; and often they do not pay to hydraulic. They are marked by a heterogeneous collection of quartz and metamorphic gravels robbed from the older streams, and by rhyolite boulders and ash, as well as intermittent solid rhyolite flows.

"The principal one of these channels was undoubtedly the Ralston Divide-Volcanoville; and up near the Goggin diggings, where it more or less conformed to the old drainage, there was a fair, if somewhat spotty, gold content. Again at the Ralston pit, where it may have caught a northerly swing of the old channel over toward Brushy Creek, before the piling up of the main lava flow, and stripped the same, there appears to have resulted some enrichment. Again, west of Volcanoville, this channel has been enriched by crossing another of the old channels. This enrichment continues all the way to Jones Hill, where there was good hydraulic ground.

"Wherever the modern rivers crossed and reconcentrated the Tertiary channels, there was little relative enrichment; but wherever they crossed the channels of the old Cretaceous system, as at Horseshoe Bar, the enrichment appears to have been the greatest."

The following is quoted from the Colfax folio of the U. S. Geological Survey:

"Auriferous Gravels.

"Neocene pre-volcanic gravels.—Along Oregon Creek several bodies of gravel are exposed, lying on flat benches sometimes less than 100 feet above the stream. The gravels at Tippecanoe are 100 feet thick and consist of quartz and chert pebbles, often imperfectly washed. They contain no volcanic rocks. The course of the Neocene stream must, as shown by bedrock relations, have followed the present Oregon Creek.

The gravel at Remargis and Gales diggings, 2 miles farther up the creek, is similar. At Tippecanoe a few acres have been hydraulicked, and some work has also been done at Gales; the gravel is here 50 feet thick and is covered with 10 feet of pipe clay. Small bodies of gravel crop near Nelson mill, and below the andesite one mile east of Plum Valley. A sharply defined channel containing little if any gravel is noted at Daneckes tunnel, 2½ miles northwest of Tippecanoe. This Neocene gulch probably drained northward.

"A junction of two important streams took place near North Columbia, and here the auriferous gravels are developed to a greater extent than at any other place. In the Smartsville quadrangle there is a large area of gravel extending from Badger Hill to the limit of the quadrangle. This is continued in this quadrangle as far east as North Bloomfield, covering about 8 square miles. There was doubtless a deep channel with slight grade running from Grizzly Hill (one mile southwest of Kennebec House) to Badger Hill, where it was joined by the steeper channel of North Bloomfield from the east. The North Columbia gravels are among the most extensive and deepest known, the depth along the center of the channel being from 400 to 500 feet. The gravel in the deepest trough, exposed at Badger Hill and Grizzly Hill, is coarse and made up largely of metamorphic rocks, while the top gravel, spread out over the benches, is fine and much more quartzose. Near the surface, and especially up toward the base of the lava flow, there are heavy masses of sand and light-colored clays.

"The gravels at North Columbia are owned chiefly by the Eureka Lake Company, their claims covering an area of 1445 acres along 2½ miles of channel. A large amount of surface work has been done and 150 feet of gravel has been washed off. The deep part of the deposit exposed at Grizzly Hill could be reached only by running long and expensive bedrock tunnels; this would have been done but for the injunctions against hydraulic mining. It is estimated that 25,000,000 cubic yards have been washed off and that 165,000,000 cubic yards remain.

   *   *   *   *   *

"Mining operations from the Derbec shaft have proved the existence of a deep channel extending for several thousand feet eastward. This is not the main North Bloomfield channel though it connects with that a short distance westward. The Derbec channel, which has a steep grade, has been mined up stream from the shaft for a distance of 7000 feet, following a curve; the width of pay gravel was from 150 to 600 feet; the height was from 8 to 16 feet from the bedrock. The gravel is coarse with many boulders, some of which are of granite. The average value per ton is $2.47. The mine was in operation from 1877 to 1893, and the production in some years reached $200,000.

"There can be but little doubt that the Derbec channel continues towards Relief. At Relief erosion has exposed a deep trough in the old bedrock and about 200 acres of auriferous gravels. The oldest gravels, as usual coarser and containing less quartz, are 60 feet deep and are covered by from 100 to 200 feet of alternating sand, fine quartz gravel and clay. Some hydraulic work was done long ago at the southern and eastern rims. For many years drifting operations only have been carried on. The Union tunnel about 2500 feet long has been driven from the southwestern side of the gravel area, and amounts up to $30,000 and $40,000 per year have been produced for a number of years. Drifting has also been done from the Blue Gravel tunnel, starting from the northeastern side of the deposit.

"For a long distance east of Relief the bedrock keeps high, and no gravel outcrops along the contact, but at Mount Zion, at Devil's Canyon, fine quartz gravel, having a thickness as great as 50 feet, crops below the North Bloomfield ditch for a distance of nearly one mile. Some little hydraulic work, as well as drifting, has been done here. Many years ago the main tunnel, running due west for 1400 feet, struck bedrock pitching west. It is probable that this gravel filled a tributary running northward and joining the Derbec channel.

"At Cherry Hill, between Shands and Mount Zion, a small body of gravel crops below the North Bloomfield ditch. A few very small areas were noted at Shands; the largest was 100 feet thick, composed of well-washed pebbles, and covered by subangular gravel. The small patches north and south of Graniteville are also partly subangular gravel. Well washed gravel crops below the andesite north of the town but is thin and irregular. A small, rapidly rising channel probably continues for some distance below the lava.

"At Snow Point and Orleans are small bodies of auriferous gravel, the bedrock rising rapidly southward. At both places the gravels have been nearly exhausted for hydraulic mining. A little drifting has also been done at Snow Point. At this place the bank is 135 feet high; the lower 15 feet contains coarse gravel, covered by 90 feet of fine sandy quartzose gravel, again overlain by 20 feet of clay. At Orleans the gravel was also largely quartzose. West of Orleans is Moore's Flat, where a considerable body of gravel is exposed. It is of the same character as at Snow Point, from 100 to 130 feet thick, and is covered by andesitic breccia. Boulders of quartz, from 2 to 6 feet in diameter, are found on the bedrock. It is estimated that 26,000,000 cubic yards have been washed off, and that perhaps 15,000,000 remain.

"At Woolsey Flat there is likewise a large body of gravel exposed. The heavy gravel up to a thickness of 100 feet, is similar in character to that just described, but it is then covered by as much as 150 feet of clay. In all of these gravel bodies the gold on the bedrock is rather coarse. But little hydraulic gravel remains at Woolsey Flat, as the thickness of the nonproductive strata is rapidly increasing. The production of these hydraulic mines, while very large, is not definitely known. None of them have been in operation since 1886.

   *   *   *   *   *

"The continuation of the Minnesota channel is found one mile south of Alleghany at Smith's Flat, somewhat higher in elevation than Chip's Flat. Here a little hydraulic work has also been done and the banks are fifty feet in height. From here the channel has been drifted through to Forest. As usual in this channel the bottom gravel is coarse and contains many flat cobbles and boulders of a bluish-white siliceous slate; also much quartz. The gold on the bedrock is coarse, and has often worked its way down some distance into the decomposed bedrock. The production of this channel has amounted to several million dollars, though it is impossible to obtain exact statistics. One of the most successfully worked claims was that of the Live Yankee, extending along 2600 feet of channel. Its production was nearly $700,000.

"A small amount of heavy gravel crops at Forest, but the channel enters the northern ridge immediately and continues in a north northeast direction. It was worked by the Bald Mountain Company, from 1872 to 1879, for a distance of about a mile, producing $150,000. The gravel was extracted to a height of 3½ feet, including one foot of bedrock. The yield per cubic yard of unbroken gravel was about $7. A shaft sunk 1800 feet from the mouth shows 215 feet of clay and sand covering 15 feet of gravel; no such heavy masses of silt are found farther down on this channel. The Bald Mountain channel was found to be cut off by a lower intervolcanic channel filled with lava, but continues beyond this to the Ruby Mine and the Downieville quadrangle.

"At Blue Tent the gravel crops extensively below the lava, filling a deep trough in the bedrock, the deepest part having the same elevation as Grizzly Hill across the canyon. The bottom gravel is coarse and cemented and is covered by over 300 feet of light-gray sand and clay, mixed with fine quartz gravel. The sand is particularly abundant and nearly barren. About 15,000,000 cubic yards have been removed and some 90,000,000 remain, much of which is barren clay and sand. The lower gravel averaged 15 cents or more per cubic yard, while the sandy top gravel contained only 2½ cents.

"On the ridge northeast of Nevada City a small but rich channel has been drifted from the east and west Harmony inclines. The gravel, which is partly subangular, is taken out to a depth of four feet. In Rock Creek below the andesite lie large masses of clay and sand, similar to the deposit of Blue Tent. Still larger accumulations are exposed at Scott's Flat and Quaker Hill. The gravel, covered with rhyolitic tuff and andesite, fills the deep trough well-exposed by Deer Creek and Greenhorn River. Along the principal channel the gravels are nearly 600 feet deep; the bench gravels surrounding the deepest trough are about 300 feet in depth. At Hunt's Hill the deepest channel exposed by mining operations is about the level of the tailings in the river. North of this point it is not visible until exposed again at Blue Tent. The evidence of the bedrock relation, and the accumulation of gravel clearly show that the deep channel is continuous from Hunt's Hill to Blue Tent. A shaft has been sunk in the old diggings at Quaker Hill and bedrock was found at an elevation of about 2650 feet. A shaft sunk in the creek at Scott's Flat struck bedrock at an elevation of about 2770 feet, the lowest bedrock not being found. At Quaker Hill the width of the channel said to pay for drifting is about 130 feet and the depth of pay gravel is from four to sixteen feet. As usual the gravel is coarse and cemented in the deep trough, while the bench gravels, several hundred feet thick, are chiefly fine quartz gravel mixed with sand.

"The yield of the top gravel rarely exceeds 6 cents per cubic yard in fine gold, while the bottom gravel may be very rich. It is estimated that near Scott's Flat 12,000,000 cubic yards have been removed, while 35,000,000 measures the amount at Quaker Hill, where the gravel banks reach a thickness of 250 feet. A vast amount of workable gravel, estimated at 140,000,000 yards, remains at Quaker Hill. At both Quaker Hill and Scott's Flat it is difficult, if not impossible, to obtain dumping ground and sufficient grade for sluices.

"Deep gravel fringed with rhyolite for three miles east of Quaker Hill represents without much doubt a tributary crossing the ridge near Galbraith's. South of this place there is about 100 feet of clay underlain by some gravel. Here some drifting has been done, both on the north and south sides. Heavy clay masses are exposed at Burrington Hill, where some hydraulic work was done long ago. The gravel of this tributary has also been hydraulicked on the north and south sides of the Quaker Hill ridge.

"High bedrock appears on the ridge three miles northeast of Quaker Hill. East of this are exposed the small Red Diamond channel on the north side of the ridge, and other channels covered with deep clay on the south side. A little work has been done on all of them. At Cooper's Mill it is said that an old incline was sunk on the rim, tracing the bedrock down to an elevation of 3500 feet. If this is correct, it would be highly remarkable, as this is considerably lower than the rim rock at any point in this lava area and it would imply the existence of a closed basin. The important Centennial channel is covered by this same lava area. Buckeye Hill is a small mass of bench gravel southeast of Quaker Hill. The gravel has been almost entirely removed.

"At Red Dog and Hawkin's Canyon near You Bet the deep channel has again been exposed and is beyond doubt continued between the two points. The gravel is similar to that of Quaker Hill. The deepest gravel has been hydraulicked only at the places mentioned, but considerable drifting by means of tunnels and inclines has been done from Niece and West's claims for 1½ miles northeast on the Steep Hollow side. The channel has very little fall, the average elevation being 2600 feet. It is estimated that 47,000,000 cubic yards of gravel have been removed, leaving over 100,000,000 yards available. Much of this would, however, be difficult to wash on account of lack of grade. Reports of yield and grade of gravel are not available, but the You Bet diggings have probably produced $3,000,000.

"The Little York gravel area contained a fragment of the old deep channel, which has been almost completely removed by hydraulic mining. The character of the gravel is similar to that of You Bet. As usual the narrow, deep channel contains a hard cemented gravel 30 or 40 feet thick, capped by as much as 350 feet of fine gravel, interstratified with some clay and sand. Large boulders of quartzite and quartz occur on the bedrock, both in the deep channel and on the benches. The yield has probably exceeded $1,000,000. The continuation of the deep channel is found at Dutch Flat and its direction is plainly marked by the small intervening gravel bodies of Missouri Hill and Eastman Hill. The principal area at Dutch Flat extends east to west for a mile; the gravel has a maximum depth of about 300 feet, the lower 150 feet consisting of coarse blue gravel, largely made up of metamorphic rock, well cemented and covered by varying thickness of finer quartz gravel, clay and sand. In the lower gravel and on the bedrock heavy boulders are plentiful. The channel has a very strong grade, in marked contrast to the level stretch below You Bet. Hydraulic work has been done chiefly at the eastern and western end, at both of which places the deep bedrock is exposed. About 90,000,000 cubic yards have been washed and a considerably less amount remains. Practically the whole extent of channel has been drifted and cemented gravel worked in stamp mills. The yield is not known but probably exceeds $3,000,000. The Polar Star gravel is said to average 11 cents per cubic yard.

"From Dutch Flat the gravel area continues southward, narrowing to a few hundred feet at Squire's Canyon and widening to from 1500 to 3000 feet near Gold Run; its southern end, overlooking the American River, is called Indiana Hill. Over a large part of this area the gravel is deep, reaching in places 300 feet and even a maximum of 400 feet.

"The surface is as usual reddish, containing many small quartz pebbles and some interstratified sand and clay. The bottom gravel in the deep trough at Indiana Hill shows 60 feet of coarse cemented blue gravel with a large proportion of metamorphic boulders; the lowest trough is here from 300 to 500 feet wide. The question whether there is a deep and continuous channel from Indiana Hill to Dutch Flat is one of much importance. Deep bedrock has been found at Jehosaphat Hill, half a mile south of Dutch Flat, having an elevation of 2877 feet, this part of the channel clearly connecting with Thompson Hill, a short distance northward. In Squire's Canyon, where the gravel area narrows down to 500 feet in width and the elevation is about 3050 feet, a shaft is stated to have been sunk to a depth of about 150 feet, striking pitching bedrock at that depth, and showing the existence here of a deep trough, having an elevation of less than 2900 feet. If this is correct, there is little doubt that a continuous deep channel exists between Indiana Hill, with an elevation of 2792 feet, and Dutch Flat, with a moderate grade of 25 feet per mile toward Indiana Hill. Bedrock has again been exposed 1200 feet farther north by the Cedar Creek tunnel, and again 2000 feet from Indiana Hill by a tunnel from Canyon Creek, run by the Gold Run Ditch and Mine Company. From the former place the bedrock was said to slope gently toward Indiana Hill. The so-called '49' shaft was sunk nearly to the bottom of the channel between Gold Run and Indiana Hill, but exact data regarding its elevation were not available. Another shaft, 75 feet deep, was sunk to the bedrock of Canyon Creek about half way between Gold Run and Dutch Flat. Extensive hydraulic mining operations were carried on at Gold Run for about ten years, in which time perhaps $3,000,000 or more were extracted. Some 84,000,000 cubic yards have been washed off, but an equal quantity, estimated at 92,000,000, remains. An area of 555 acres has been washed off to an average depth of 75 feet. At Indiana Hill the bottom gravel was drifted and crushed in mills. The yield per cubic yard of hydraulic gravel is said to be 11 cents, but this estimate is in all probability too high. The drifting ground at Indiana Hill yielded between 1872 and 1874 at the rate of $9 per cubic yard of gravel in place. Above Dutch Flat toward Alta is the gravel hill of Narv Red, the narrow channel of which has been drifted and hydraulicked; the gravel is a medium, fine red quartz, covered with rhyolitic clays. From here the channel extended in the hill toward Alta. A shaft sunk at Alta 35 feet below the railroad found bedrock at 132 feet. A tunnel extends from Canyon Creek one-half mile south of Alta to the shaft and the gravel in the channel is now being worked. The gravel is soft quartzose, not cemented. From this point a branch channel probably crosses Canyon Creek and extends to Moody Gap, east of which the remainder is probably eroded. Another branch extends from Alta eastward, probably emerging at Shady Run, and grading sharply westward. It is mostly filled with rhyolitic clays, although a bank of gravel also appears on the northern rim which has been washed. Minor drifting operations have also been undertaken in this vicinity. A remainder of the same channel is preserved at Lost Camp two miles south southeast of Blue Canyon. Here are about 120 acres of quartzose, imperfectly washed gravel, 50 to 75 feet deep, containing some rather large boulders. Only a smaller portion has been hydraulicked.

"A branch of the Dutch Flat channel continued across the present Bear River. Elmore Hill on the point between Bear River and Little Bear Creek has been almost completely washed off. Rising at a rapid rate, the continuation of the channel is found at Liberty Hill. The gravel is here about 60 feet deep, 30 feet of reddish quartz gravel covering the same amount of blue gravel full of very large boulders of gabbro and serpentine. The amount of gravel removed is estimated at 2,000,000 cubic yards, some 16,000,000 remaining. The channel continues up to Lowell Hill, but the gravel is here covered by very heavy masses of light-colored clay. At Lowell Hill the gravel is 30 feet deep, the coarse bottom gravel being covered by finer quartzose gravel. The heavy clay banks make hydraulic working difficult. Considerable work has, however, been done at the Klamath Mine. Drifting operations have also been undertaken south of Nigger Jack Hill at the Valentine Mine and farther south opposite the Klamath at the Swamp Angel.

"Opposite Lowell Hill lies Remington Hill at a slightly higher elevation. Here again is an old depression filled with gravel of which a few acres are exposed. The gravel is similar to that of Lowell Hill and is capped by heavy masses of clay. The amount excavated is estimated at 1,750,000 cubic yards, while possibly 600,000 cubic yards remain. Much of this, however, is heavily capped by clay and volcanic tuff. The channel has been struck by two drift tunnels a little eastward, making it possible that the channel comes out again at Democrat, another little gravel point, separated from Remington by a bedrock spur, where hydraulic work has also been done.

"On the point between the forks of Steep Hollow, opposite Democrat, is the small gravel hill called Excelsior, doubtless representing the extension of the Democrat channel. To the north and northwest of Excelsior the bedrock rises rapidly. The channel may have continued a couple of miles farther northeast, but whether it enters under the lava flow or follows the present course of Steep Hollow is uncertain.

"On the south fork of the Yuba several important gravel bodies are found. A few small points covered with quartz gravel occur southeast of Relief on the south side of the canyon. At Alpha about 75 acres of gravel are preserved, the pebbles consisting chiefly of quartz, quartzite and a hard conglomerate. Some quartz boulders on the bedrock reach 5 feet in diameter, but most of the gravel is light and sandy. The banks are 90 feet high, including 20 feet of clay at the top. The amount removed is 5,000,000 cubic yards; only a quarter of that amount remains.

"At Omega several hundred acres of gravel are exposed and have been extensively worked. The gravel lies on a flat bench and apparently extends southeasterly under the lava. The greatest thickness is 175 feet. The bed consists of 150 feet of gravel, covered by six feet of clay, above which is again 20 feet of gravel, all showing colors. The lowest stratum contains some large boulders of granite from the Canyon Creek area, but the main body is composed of smaller cobbles up to six inches in diameter, quartz decidedly predominating. The extent of this channel southward is not definitely known, though a shaft was sunk to bedrock on the Blue Tent ditch, cutting good gravel. Its depth is not known. Some distance south of Omega is a small gravel flat called Shellback at a higher elevation; beyond this the bedrock rises rapidly. Towards the southeast the bedrock also rises, though less rapidly, and gravel is found in places along the rim. At Diamond Creek a small body of quartz gravel is exposed, having a maximum thickness of 12 feet, and covered by nearly barren Pleistocene morainal boulder clay.

"Extensive hydraulic operations have removed 12,000,000 cubic yards at Omega, the tailings discharging in Scotchman Creek through a 3000-foot bedrock tunnel. Apparently reliable calculations give 13½ cents as the yield per cubic yard; the lowest gravel of course being much the richest part of the deposit. About 40,000,000 cubic yards are estimated to be still available for hydraulic mining.

"It remains to mention the occurrence of many uncertain and puzzling features at Phelp's Hill, Centennial, and San Jose shafts. At Phelp's Hill, at an elevation of about 4060 feet, 15 to 30 feet of gravel outcrop below the lava for one-half mile. Heavy quartzose boulders are found on the bedrock. A remarkable disturbance occurs here, the gravel being cut by a fault which throws the west side down about 40 feet. The fault is traceable for at least 400 feet, running north and south. The Centennial shaft, one and one-third miles south southeast of Phelp's Hill, was sunk in 1887 to a depth of 400 feet, and the bottom of a deep channel was found by drifting from it. Later a tunnel was run 2500 feet south from the place indicated south of Phelp's, the elevation being about 4080 feet. A channel was struck at the tunnel level; it is 400 feet in width, and carried quartz and greenstone gravel, the gold being fairly coarse. Work has been suspended, from which it may be inferred that on account of its width the gravel body on the bedrock is not very rich. If, as seems probable, this channel connects with that of Phelp's, it can have but little grade.

"A mile southwest from the Centennial shaft the San Jose shaft is sunk in the bed of south fork of Deer Creek to a depth of 340 feet, giving the channel an elevation of between 4000 and 4100 feet, which is stated to be somewhat higher than the Centennial channel. Drifting from the shaft showed the channel to be about 300 feet wide. The gravel is composed of cobbles of quartz and country rock about seven to fifteen feet thick, covered by 40 feet of clay, above which is lava. There is little doubt that this channel is continuous with the Centennial, and it appears probable that its grade is northward, making it a branch by way of Phelp's Hill of the main stream from Relief Hill to Omega. It has been thought by some that this channel might continue to Remington Hill with a southerly grade. This appears unlikely, however, and it is scarcely possible that there should be a continuous channel between Pheln's and Remington hills, for the channels at these two places certainly connect with different branches of the old Yuba River. There will probably be found a low divide separating the San Jose channel from Remington Hill and from the Quaker Hill drainage. It is also very unlikely that any of the channels under this lava area had any direct connection with Omega.

"On the Iowa Hill and Forest Hill divides a small amount of gravel is exposed on the surface, but the channels preserved below the lava are rich and numerous. At Iowa Hill a deep channel extends from northwest to southwest across the ridge north of Indian Creek. The sharply defined trough is 200 feet deep and is filled with coarse gravel, well cemented in its lower parts. The total thickness is over 300 feet. The channel is from 200 to 400 feet wide on the bottom. This gravel has been hydraulicked, except a narrow ridge upon which the town stands. Lighter quartzose bench gravels extend northeast of Iowa Hill. They have a maximum thickness of 200 feet and are covered by thin rhyolite tuff and andesite. They have been extensively hydraulicked and some ground yet remains.

"At Succor Flat a deep and narrow channel, belonging to the intervolcanic epoch, has been drifted for a distance of 2500 feet. The same channel probably crosses Indian Creek at Monona Flat and finds its outlet at some place on Roach Hill. South of Indian Creek over 300 feet of gravels crop; southward they thin out with rising bedrock but deepen again near Wisconsin Hill, having at both places the same general character as at Iowa Hill. Between Morning Star and Wisconsin Hill there is doubtless a deep and continuous channel, which is clearly the extension of that underlying Iowa Hill. Extensive hydraulic work has been done, both near Morning Star and east of it along Indian Creek, as well as at Wisconsin Hill. A body of higher bench gravel, across Refuge Canyon at Elizabeth Hill, has also been hydraulicked, but nearly all of this work has ceased during the last decade. Instead extensive drift mining has been carried on. At the Morning Star the deep channel, extending in an easterly direction, has been mined for a distance of nearly 3000 feet; about seven feet of cemented gravel are extracted, the width of the pay gravel being from 80 to 200 feet. This drift mine has proved among the richest in the gold belt. The gravel contained for a long period, it is stated, $7 per carload, equal to $14 per cubic yard, and the annual production ranged from $2,500 to $150,000.

"The Big Dipper Mine has been working the same channel since 1890 from the Wisconsin Hill side with excellent results. The grade of the main channel is remarkably slight, 2692 feet being the elevation of bedrock at Wisconsin Hill, 2685 feet at the Morning Star, and 2631 at the Northwest side of the Iowa Hill channel. In 1899 the workings of this mine were connected with those of the Morning Star, proving conclusively the identity of the channels. A smaller channel pitching into the ridge has been followed some distance in from Grizzly Flat and probably joins the Morning Star channel. A small body of well worked quartz gravel was found at Kings Hill, chiefly one and one-half miles southwest of Wisconsin Hill; it is interesting because of its position between Yankee Jim and Wisconsin Hill, and its comparatively low elevation—2550 feet. Four or five acres have been washed here to a depth of 20 feet.

"Above Monona Flat very little gravel is exposed, the andesite tuff resting on bedrock of irregular configuration. At the Giant Gap claim, four miles west of Damascus, the lava cap is very narrow; below it a gorge-like intervolcanic channel has been exposed. Three miles west of Damascus is McIntyre's claim, where a thousand foot tunnel has exposed the same or a similar narrow channel. One mile northeast of this is the Colfax claim, showing some quartz gravel, probably belonging to a prevolcanic channel, a continuation of which may be found at Jimtown, three-fourths of a mile north of the reservoir. At Jimtown a shaft, 100 feet deep, has been sunk, finding quartz gravel and pitching bedrock. No data are available to estimate the yield of Iowa Hill Divide since 1849. It probably considerably exceeds $10,000,000.

"To begin now a rapid sketch of the Forest Hill divide, it should be stated that comparatively little of the mining work done falls south of the boundary of the Colfax quadrangle. At Peckham Hill a little unsuccessful drifting has been done on the deep and narrow cement channel finding its outlet there. At Todd Valley a body of bench gravel crops, which was washed at Pond's claims until the overlying lava became too heavy to handle. This gravel is partly cemented, poorly washed, and about 40 feet thick. About 11,000,000 cubic yards have been washed off, the yield of which is given as $5,000,000, but this is probably too high.

"At Georgia Hill, opposite Yankee Jim, a thickness of 100 feet of gravel is exposed below the lava, and a few acres have been washed off along the edge. At Yankee Jim a larger area of gravel from 40 to 100 feet thick is met with, which toward the east disappears under the lava. The gravel is fairly coarse, being composed of metamorphic rocks with some quartz. The bedrock is at nearly the same elevation as at Georgia Hill, and the main channel seems to have had its direction northeasterly and southwesterly, though a somewhat higher channel extended eastward, and probably connected with the Smith's Point bench gravel a mile and a half distant and situated on the south fork of Brushy Creek. The gravel at Smith's point is fifty feet thick, interstratified with sand. It is estimated that 8,630,000 cubic yards have been removed from Georgia Hill, Yankee Jim and Smith's Point, and that the yield has been about $5,000,000. The amount remaining available for hydraulic work is undoubtedly less than that removed, for the volcanic cap will soon make hydraulic work impossible. One-quarter of a mile east of Georgia Hill the Anthony Clark tunnel has recently been run in a southerly direction about 550 feet, and it is reported to have shown the existence of a large channel with much granitic detritus. The tunnel is found to be too high, striking the channel above bedrock.

"It is believed that the Yankee Jim channel flowed northward toward Wisconsin Hill by way of Kings Hill. It is also believed that it connects below the lava with the Dardanelles channel, though the latter intervolcanic channels may have removed much of the earlier accumulations and in some places destroyed the older channel.

"At Dardanelles and Forest Hill the canyon slope has exposed below the lava a long, low trough, filled with gravel and rhyolitic tuff. The gravel is moderately coarse, composed of quartz and metamorphic rocks, and is well cemented near the bedrock. Above it rests rhyolitic tuff, intercalated with some gravel, clay and sand. The thickness of these two formations varies exceedingly. At the New Jersey claim the gravel is only eight feet thick, and is overlain by rhyolitic tuff. At the Dardanelles it has a maximum thickness of 70 feet. In the region above Mayflower are extensive bodies of rhyolitic tuff with intercalated gravels, as well as clays and sands, of more doubtful origin. The depth of these accumulations at Mayflower over the deep channel is 350 feet. In the intercalated gravels granitic and rhyolitic cobbles are common. At Adams tunnel 178 feet of rhyolitic clay are exposed with two smaller gravel bodies. Again at Black Hawk, Wasson, and Westchester claims similar bodies are exposed. At Bath, again, the same channel is exposed with about 250 feet of overlying gravel and white tuff. The lower part is a trough 500 feet wide and 100 feet deep, filled in the bottom with washed and rounded bedrock boulders, composed chiefly of serpentine and greenstone. Above this comes a thick stratum of the usual coarse quartz gravel, and above this a thick series of rhyolitic tuff with intercalated gravel, having a maximum thickness of 30 feet, and containing granite and rhyolitic boulders. The thickness of this series varies from 100 to 250 feet, and it is again covered by 270 to 300 feet of andesitic tuff breccia.

"The main prevolcanic channel enters the ridge at Bath and runs northerly for a mile with very slight grade, then curves west and south, assumes a grade of 60 feet per mile, and passes below Mayflower and Forest Hill to the Dardanelles, where it turns northwest again towards Yankee Jim without leaving the ridge.

"The mining operations in this vicinity have been very extensive. The hydraulic operations have mainly ceased, though a considerable amount of ground is still available at the Dardanelles and around the head of Brushy Canyon. At the former place and at Forest Hill 4,850,000 cubic yards have been excavated; at the head of Brushy Canyon probably 7,350,000 cubic yards. Only drift mining is now carried on.

"The main old channel has been drifted at Dardanelles for 2500 feet in a northwest direction; the gravel which is cemented was here five feet deep and 75 feet wide. Mining is still in progress here. The mine is believed to have produced $2,000,000 or more by drifting and hydraulicking.

"Below Forest Hill a number of smaller depressions, called front channels, were worked many years ago from Jenny Lind and New Jersey tunnels, but no extensive recent work has been done there. The main channel has been reached by the Baltimore tunnel and Excelsior slope, but some drifting ground still remains between these points and the Mayflower. The ground in this vicinity is supposed to have produced $5,000,000, about $1,500,000 being taken from a strip of ground in the New Jersey claim, 800 feet long and 300 feet wide.

"From the Mayflower tunnel, 4740 feet long, the main channel has been worked, chiefly from 1888 to 1894, for a distance of three miles, connecting it with the Paragon workings. A bed of gravel from two to fourteen feet thick, having an average width of 75 feet, was removed from the bedrock. The yield has been approximately $1,500,000, or $7 per ton of loose gravel delivered. Sixty-six per cent of the bottom gravel was found to pay for extracting. Between the Paragon and the Mayflower in the bend is a narrow gorge, 1000 feet long, where the channel is only 25 feet wide and poor in gold. An upper lead or streak of gravel enclosed in the rhyolitic tuff, 150 feet above the bedrock and paying for drifting, is said to exist along the Mayflower channel, as well as at the Paragon at Bath, but it has not yet been worked to any extent. Little work is being done at present on the main channel at the Mayflower. The same channel has been worked from the Paragon mine to a distance of 6800 feet north. The width of gravel breasted is 50 feet, depth 2 to 7 feet, yield per ton delivered at surface $10, total yield by hydraulicking $500,000, by drifting, $850,000. At the Paragon there exists an upper streak of pay gravel 150 feet above the bedrock. This was followed for 2000 feet until cut off by a channel of intervolcanic erosion filled with andesitic tuff. The width of this upper lead was 225 feet, the depth of non-cemented pay gravel 5 feet, and the yield per ton of loose gravel $4.50. The total yield was $900,000. The mine has been operated for 36 years, and the channels are said to be nearly worked out.

"A portion of what is doubtless the same channel has been preserved at Michigan Bluff. The deposit which covers about 40 acres is composed of pure quartz gravel; on the bedrock lie huge rounded quartz boulders. Some 6,000,000 cubic yards have been removed and a smaller quantity remains. The yield is reported to have been $5,000,000, some of the ground being exceedingly rich. The deposit bears the character of bench gravel. At Sage Hill and Birds Valley a long, narrow channel, with strong southwest grade, is preserved. The outlet of it at Sage Hill is somewhat lower than Michigan Bluff. It has been worked to some extent, but is not so rich as that at Michigan Bluff. Much coarse, rough and crystallized gold was found here, as well as in Mad and Lady canyons.

"At Edwards Hill a small patch of partly volcanic gravel has been worked. From here north a number of small gravel points appear, most of which belong to intervolcanic channels. At Gas Hill, however, there is a patch of the same quartz gravel as is exposed at Michigan Bluff. Immediately to the north it is eroded by deeper volcanic channels, but between Hidden Treasure and Damascus a nearly continuous old prevolcanic channel, having a grade of 70 feet per mile southward, has been found under the lava cap. This is a wide, flat channel, filled with about 200 feet of non-cemented quartz gravel, sand and clay. The material is decidedly finer than that of the Bath-Mayflower channel, although some quartz boulders may be found on the bedrock. It is cut off by two deeper intervolcanic channels, one a mile south of Damascus; another one and a half miles north of Sunny South. Between these a fragment of the old white channel remains. This channel was first found at Damascus and drifted on until cut off by the intervolcanic channel mentioned. The yield of this part is reported to be $6,000,000. From Sunny South the Hidden Treasure mine has worked the deposit 7700 feet northward, width of gravel breasted 250 feet, depth 4 to 7 feet, yield of loose gravel delivered $1.75 per ton. The total yield to 1890 was $1,150,000, and up to 1898 probably nearly $2,000,000. Since that time the operations at Sunny South have been discontinued, and another tunnel has been started at the Dam claim, one mile farther north, from which the fragment of channel remaining between the volcanic channels is not being mined. The mine has been worked for 23 years.

*   *   *   *   *   *   *

"The general Neocene drainage of this quadrangle has been roughly considered under the heading of auriferous gravels, but it remains to indicate in a more detailed way the connection of the channels of the southern part of the area with those of the region between Dutch Flat and North Columbia.

"There is not the slightest doubt that a river, corresponding roughly to the present middle fork of the American, had its source near Castle Peak, thence flowed across to Soda Springs, and approximately followed the present middle fork, entering this quadrangle under the ridge between Long Canyon and the middle fork, and in the southern portion of this ridge curving into the Placerville quadrangle. It entered the Colfax quadrangle again a few miles west of this, and the channel emerged from under the volcanic capping at Ralstons. A tributary from the Duncan Peak region joined it with a general southerly direction. From Ralstons much of it eroded, but it may be regarded as certain that the main channel continued westward, touching Michigan Bluff and Sage Hill, here receiving an important tributary, running nearly due south from Damascus. The deposits of this latter channel are preserved below the lava ridge, between Damascus and Gas Hill. Near the latter point it receives a tributary from Last Chance and Deadwood.

"Again east of Michigan Bluff the channel is eroded, but it is certain that its continuation is found at Bath, whence the main channel ran through the Mayflower. Here it made a wide curve and ran southward to Forest Hill and the Dardanelles. Thus far the general course is outlined without uncertainty, but from here on the difficulty begins. This main channel is marked by its heavy deposits of gravel and clay, and its broad well-defined channel. Under the southwestern prolongation of the Forest Hill lava ridge nothing has thus far been found which would indicate that the main old river channel flowed down in this direction. It is true that a narrow channel of the intervolcanic epoch continues down in this direction, but these channels were notably independent of the older and main drainage basins. The intervolcanic channels were excavated after a large part of the old river basin had been filled by accumulations of silt and volcanic mud, and probably also after the tilting of the Sierra Nevadas had taken place. Their direction, then, offers no criterion of the prevolcanic drainage lines. It would certainly seem as if some fragments of the accumulations of the old channels would have been preserved southwest of the Dardanelles had the channel taken this course. The gravels exposed at Todd Valley offer no solution of the problem, for they are at a higher level and evidently represent a bench filled with gravel after the clogging of the main channel.

"There is, however, a solution of this problem, which is advanced as having many plausible points, though it can not be said to be free from all objections. This is, that the old channel of the Forest Hill divide emerges at Yankee Jim and Georgia Hill, and that its course from there is northward to Wisconsin Hill, thence through the lava ridge and curving eastward to the Morning Star mine, thence to Iowa Hill, crossing the canyon of the present river to Indiana Hill and from there northward to Dutch Flat, whence its course has already been established. This hypothesis, in the first place, necessitates the existence of a deep and continuous channel between Dutch Flat and Indiana Hill. That such a deep channel exists appears now very probable and may be regarded as certain if the development south of Dutch Flat and Squires Canyon will show the existence of a deep trough at this place, which it has been asserted was found by the exploration. One of the principal difficulties appears to be the fact that the gravel at Georgia Hill and Yankee Jim differs somewhat in character from that of the Mayflower and Forest Hill. This may be explained by the fact that the river at this point spreads over a larger and flatter bottom, which would naturally influence the character of its deposit.

"The difficulty, which at first glance appears to be an insuperable one, that is, that of the grades, on closer examination, converts itself into an argument in favor of this hypothesis. From the Dardanelles to the Yankee Jim is a slight grade which is sufficient for the requirements. From Yankee Jim to Wisconsin Hill the channel would at present have a slight upward grade. From Wisconsin Hill to Iowa Hill it is apparently approximately level. From Iowa Hill to Indiana Hill it has a slight southward grade and similarly from Dutch Flat to Indiana Hill is a grade which, though slight, is opposite to that which the river, according to this hypothesis, would have had.

"From Yankee Jim to Dutch Flat the Neocene river would have pursued a nearly due northerly course; now it is likely that this river from Yankee Jim to Dutch Flat had originally a very slight grade northward, similar to that of the Neocene river between You Bet and North Columbia. Examinations of channels in various parts of the Sierra Nevadas have shown the occurrence of a tilting movement which has affected the grades of the channels according to their direction. Channels running north northwest to south southeast would retain their original slight grade. Those running west of this line would have their grades materially increased by the tilting. On the other hand, those flowing in a more or less easterly direction from this axis of tilting would have their grade decreased or even reversed. A close examination of the elevations of Indiana Hill, Dutch Flat, Iowa Hill, Wisconsin Hill, and Yankee Jim will show that, in fact, the present level character or slight southward grade of these channels is exactly what would follow if the Neocene river with a northerly course had participated in a westward tilting of the block of the Sierra Nevadas, amounting to about 60 or 70 feet per mile.

"If this hypothesis be true, it solves in an exceedingly satisfactory way, a number of the perplexing problems which were presented by the enormous accumulation of gravels in the drainage of the old Yuba River. It increases vastly the watershed of the Neocene stream, which as now outlined extends from the headwaters of the north fork of the Yuba. The waters of all this territory found an outlet through the narrow channel from North San Juan to Smartsville. In the central part of this drainage area longitudinal depressions existed, bordered on the west by the high diabase ridges of the foothills. All these conditions naturally tended greatly to increase the accumulation of gravel. What has formerly been supposed to be the north fork of the Neocene American River now becomes the south fork of the great Neocene Yuba River. The Neocene American River is reduced in size, and consists only of the stream coming down from Pyramid Peak by way of Placerville.

"During a rather long interval between rhyolitic and andesitic flows new channel courses were established. A disturbance had taken place that increased the slope of the Sierra Nevadas and the streams began active cutting; thus on the Forest Hill divide there exists a complicated system of narrow, deep channels, which in many places have destroyed the old ones. These intervolcanic channels, often called cement channels, belong to at least two systems, the younger being characterized by a large amount of coarse volcanic gravel, rarely containing much gold, and having been formed after the first andesitic flows had already invaded this region. The older system carries thin, mixed metamorphic and volcanic gravel, rarely more than ten feet thick, there being no gravel at all along certain parts of the streams. This gravel lies on the naked bedrock, and is covered by a series of flows of andesitic tuff, the lowest usually fine-grained, and referred to as chocolate or cement; the upper flow consists of the usual tuffaceous breccia. Strata of gravel and sand of mixed character, volcanic and metamorphic, are often found interbedded with the andesitic tuff. Wherever the intervolcanic channels have robbed the old channels, they are likely to be rich, though irregular as to their pay. Some of them, however, have been found unexpectedly poor. The gold is usually coarse. The upper gravels in the andesitic tuff sometimes carry gold, though seldom enough to pay for drifting. Some of the volcanic channels have not only cut through the old channels but have eroded small canyons in the bedrock up to a depth of 150 feet. One of the most conspicuous of these crosses Volcano Canyon, and is exposed by the Hazard shaft. The grade of these channels is always steep, usually from 70 feet per mile upward.

"A whole channel system, belonging to this period, is buried below the lava of Forest Hill divide. The principal channel can be traced almost continuously from the Weske tunnel, above Michigan Bluff, down to the outlet of Peckham Hill. It cuts the old channel several times, and receives numerous tributaries, preserving throughout the same character of a deep erosion channel, sometimes barely reaching the bedrock, sometimes cutting deeply into it.

"At Peckham Hill and Blue Gravel shaft, in the Placerville quadrangle, it has been opened but apparently does not pay. For 2½ miles north of Peckham Hill it has not been bottomed, but at Gray Eagle tunnel it has been opened by a tunnel from Owl Creek, 2500 feet long, and a shaft 360 feet deep. Though somewhat too high, the tunnel has followed the channel up stream for several thousand feet. The pay is spotted, the gravel thin, though often rich. In the Mayflower mine, the channel is again exposed; it is here called the Orono, and has cut down to about the level of the bedrock in the Mayflower channel. From here it has recently been worked for a distance of 2000 feet through the Mayflower tunnel. Again, a little below the mouth of the Mayflower tunnel, in Brushy Canyon, a channel crosses the canyon at a lower elevation than the

Mayflower, called the Live Oak. It has been drifted upon northward for 2000 feet. Southward it probably joins the Orono channel. Below the volcanic capping, between the forks of Brushy Canyon, are several intervolcanic channels, such as the Adams, Nil Desperandum, Westchester, Black Hawk, and Wassen, the relations of which are little known.

"Farther east the main channel is again found in the Paragon mine, where it has not quite cut down to the bottom of the old channel. Again it is exposed where it crosses Volcano Canyon, in which the Hazard shaft has been sunk 180 feet; the narrow channel was followed west for 3000 feet and some rich gravel was found. Above there are about two miles in which the channel has not been exposed, though a deep tunnel from near Michigan Bluff has been proposed, but above this, it has been drifted for over 5000 feet in a westerly direction from the Weske tunnel. In spite of difficult working conditions, this enterprise yielded excellent returns, producing approximately $750,000.

"A smaller intervolcanic channel, filled with heavy volcanic gravel, crosses the Weske channel near its inlet and thence continues some distance north. It has not been worked to any extent. About a mile north of the Weske channel a small old stream bed has been worked to some extent from the Bowen and Oro tunnels. The latter is about 2500 feet long.

"Above Weske tunnel, confronting El Dorado canyon, there are a number of smaller gravel hills, most of which have been hydraulicked. Among these are Drummonds Point, El Dorado Hill, and Batchelor Hill. The gravel at all of these places appears to belong to the intervolcanic epoch, and the deposits evidently form part of a somewhat complicated channel system near the point where the channels from Deadwood join those coming down the main ridge. It is probable that the channel on which the Oro tunnel is driven finds its way down below the level ridge on the western side of the Hidden Treasure tunnel, but it has not been exposed north of the tunnel mentioned.

"A narrow, intervolcanic channel, with heavy volcanic gravel and apparently barren, runs north for some distance from Sunny South, parallel but a little east of the Hidden Treasure channel. At Sunny South it has cut across the latter, obliterating it and eroding some distance into the bedrock below the level of the Hidden Treasure. This is the reason why no quartz gravel can be seen cropping out at Sunny South. About a mile south of Damascus the Mountain Gate channel was cut off by a deeper, intervolcanic water course, eroded to a depth of about 150 feet below the older channel. This so-called Blue channel was drifted from the Mountain Gate tunnel, producing $175,000. A little over two miles north of Sunny South the same old channel is cut to about the same depth by another intervolcanic channel, finding its way southeasterly to the Dam claim, and thence for a mile farther in the same direction to the Mitchell claim. The Dam channel, though narrow and irregular, has been drifted for 2500 feet northwest of the point where it crosses El Dorado Canyon. The Mitchell claim on the same channel has also been worked for a distance of 2000 feet. Still another intervolcanic channel, called Bob Lewis channel, has been worked for 1000 feet, south of its inlet on the east side of the Mountain View channel at Damascus. The principal intervolcanic channel, which probably continues from the Oro to the Blue channel of the Mountain Gate tunnel, has again been exposed at Red Point, and worked for a distance of 12,000 feet up stream from the Red Point tunnel, which strikes the channel 2000 feet from its mouth. The Red Point channel is somewhat irregular in width and depth of gravel and in pay. The average fall of the channel is 75 feet per mile. The width of gravel breasted is 120 feet, the depth from 2 to 12 feet but generally small. As delivered at the surface, the gravel contains $2.50 per ton. Volcanic pebbles are of common occurrence in the wash material. The Red Point mine has been worked for ten years, and during that time has been a steady producer. It is immediately capped by the hard, andesitic tuff. Large wash boulders, often two or three feet in diameter, occur in the gravel. The total production during the five years from 1888 to 1892 was $368,000, and it is believed that since that time an almost equal quantity has been extracted.

"As we approach the higher region of the Sierra, where accumulations of prevolcanic gravel were small or did not exist at all, the difficulty of distinguishing between prevolcanic and intervolcanic channels becomes greater. Strictly speaking, all of the channels must be considered as belonging to the later group, as some erosion necessarily took place in all of them in which bedrock was exposed. Going up toward Duncan Peak we find in general that the grades of the channels increase and that they assume more and more the character of narrow tributaries or gulches.

"It is believed that the Red Point channel continues up the ridge. It has indeed been exposed at the Hogsback tunnel 5½ miles northeast from Red Point. The tunnel runs south-southwest 2500 feet, exposing a very deep and narrow gorge with steep westerly grade; and contains very little gravel. Though yielding some gold, the channel was not found to pay. About a mile south of the Hogsback channel another deep ravine has been exposed at the Greek mine and the Black Canyon, between which points it is probably continuous. The Black Canyon has been worked for 700 feet eastward. The channel is narrow and very steep, having a grade of 7 feet per 100, with several abrupt falls. On the bedrock rests a few feet of coarse gravel, containing very coarse gold. Above this lies 50 feet of andesitic tuff, gravel, and sand interstratified. No volcanic pebbles were seen in the gravel and the channel probably belongs to the prevolcanic period. The cost of working this channel is necessarily very high. The inlet of the Hogsback channel is probably found at the low place half a mile north of Secret Canyon House.

"Near Canada Hill another steep, narrow channel has been exposed which appears to have a very sharp northeasterly grade and the direction indicated on the map. This channel probably crosses Sailor Canyon, entering the Truckee quadrangle, and then joins the main channel, following approximately the Middle Fork of the American somewhere near French Meadows. The western end of the Canada Hill channel is not covered by volcanic rocks but by heavy morainal detritus. A short distance eastward the volcanic rocks begin and cover it to a depth of about 100 feet at the Reed mine, a half mile east of its beginning. A few feet of poorly washed gravel are found in the bottom of the channel, above which are a few feet of clay containing carbonized wood. Above this lies a little massive rhyolite covered by heavy masses of andesitic breccia. This channel has been successfully drifted and in places hydrau-licked as far as the place where it enters the high volcanic ridge. It is believed to continue with steep grade underneath this ridge, and its outlet has probably been found at the Sailor Canyon mine two miles northeast of Canada Hill. At this place bedrock tunnels have shown the existence of a narrow channel containing angular, poorly washed gravel covered with a dark clay. The relations at this place are some-what obscured by considerable masses of morainal material.

"Deadwood Ridge is crossed by channels belonging to both the earlier and later periods, which have been extensively worked. The older channel is believed to enter the ridge somewhat south of the Devils Basin, and finds its outlet half a mile north of Deadwood. It is characterized by thicker gravel bodies containing large boulders of quartz and metamorphic rocks. This channel has been worked from the Rattle-snake mine on the eastern side of the ridge, and from Reed's and Hornbush's tunnels on the western side. The principal intervolcanic channel has its inlet at the Devils Basin, and has been worked from there for a distance of half a mile, yielding very rich returns. The thickness of the gravel is said to average two and a half feet. The outlet of this channel is probably 3000 feet north of Deadwood, and somewhat lower than the adjoining outlet of the older channel. From this side it has been worked 3000 feet eastward without, however, connecting with the basin tunnel. A second intervolcanic channel enters the ridge south of Deadwood and runs in a northerly direction. It has been followed down stream for 3000 feet.

"At Last Chance several channels are known to occur and have been drifted for a considerable distance, although some ground is yet unopened. As at Deadwood there is a prevolcanic channel and several intervolcanic channels. Both classes follow approximately the same course, though the intervolcanic channels are about 20 feet lower than the others. The gravel and its covering material are similar in character to that of Deadwood. The upper continuation of the Last Chance channel may prob-ably be found at American Hill on the ridge between Lost Canyon and Antone Canyon.

"Below the volcanic areas, south of Duncan Peak, narrow and deep channels have been found which, however, have not thus far yielded much. One of these extends from Flat Ravine southward for 1½ miles. It has been opened by tunnels at both ends and worked to some extent. Another channel is exposed by the Abrams tunnel on the west side of Duncan Canyon. This branch probably joins that from Flat Ravine and, crossing under the lava ridge between Duncan Canyon and the middle fork of the American, becomes a tributary of the main Long Canyon channel. Depressions indi-cating channels also exist below the andesite areas of Big Oak Flat."

In the above described area there is still a great deal of virgin ground on which some drift mines may possibly be developed. The course of the old Breece and Wheeler channel, where it goes through the lava caps of Nevada Point, of Devils Peak and of the Ralston Divide, offers an opportunity for some profitable drift mines. The Glenn Mine, on the channel coming down from Flat Ravine, has produced considerable money in the course of the last two years, and it is quite possible that some of the other tributary channels may be well worth developing. In commencing drift operations the primary distinction to be made is that between the prevolcanic or Cretaceous channels and the inter-volcanic or Tertiary channels, which latter are of little importance from an economic standpoint, except where they may have robbed gold from the older channels.

There remains to be spoken of, in connection with this channel system, a tributary on the north side of the main ancient river which comes in near Badger Hill and can be traced up the ridge between Oregon Creek and the Middle Fork of the present Yuba. This channel has been worked at Grizzly Gulch, at Tippecanoe and at Nigger Tent, and possibly contains some ground yet which would pay to drift.

In addition to this, we have the Nevada City and the Grass Valley channel systems which are more or less worked out, but which have produced a great deal of gold.

Coming back to the Nevada City and Grass Valley region, we have a large tributary entering the main channel at or near Mooney Flat above Smartsville. This channel has been eroded along Negro Creek, but appears as hydraulic diggings on the south side of the ridge between Negro Creek and Deer Creek. From here on it can be traced through Randolph Flat, where it is joined by a tributary from the southeast, known as the Old Alta Hill channel. This channel can be traced directly under the town of Grass Valley and southeasterly by way of Osborne Hill. Another channel, coming down from Kres across the head of Little Greenhorn Creek, joins this channel at or near Grass Valley. Still another one comes down from Banner Hill in a southwesterly direction. A westward flowing channel from Crystal Spring comes through the Towntalk Ridge and joins this channel near Randolph Flat. Still another from the Harmony Ridge, passing through Nevada City in a southwesterly direction, undoubtedly joined this same system at or near Randolph Flat. On the other hand, a large portion of this drainage from Nevada City undoubtedly flowed northerly and westerly through Montezuma Hill; thence southwesterly through Bunker Hill, across the south fork of the Yuba River in the general direction of Kentucky Ravine. Still another cement channel runs across through Shelby toward Jones Bar.

Practically all of the Grass Valley and Nevada City systems is now of little economic importance, as the richest of the drift ground has been worked out, and only the Tertiary and cement channels remain virgin. The accompanying map shows the general distribution of these systems.

The area described in what is known as the Colfax quadrangle probably contains the most important bodies of hydraulic gravel yet remaining in the State of California. In the Downieville quadrangle to the north, there are large areas of higher grade gravel, but they are not nearly so extensive.

It has already been mentioned that in this area evidence of the glacial origin of the Cretaceous white gravel channels is so clearly manifest that it can not be overlooked. A sketch or profile showing the elevation of the Tertiary Yuba from Badger Hill to Michigan Bluff is inclosed. The data for this profile were taken from Professional Paper 73 and maps of the U. S. Geological Survey, supplemented by detailed private surveys and other data. The projected fault corresponds to the North Bloomfield fault. The diverging of the Tertiary South Yuba and American channels from a common source and the occurrence of diverging runs with apparently different courses in the same channel system can be paralleled among the glacial channels in Alaska.

The top gradients of the gravel channels are such as prevail on the debris deposits resulting from operating the hydraulic mines.

This profile was prepared to accompany a paper upon the origin and deposition of the gravel deposits, which is now in preparation by Mr. W. W. Waggoner of Nevada City.

It will be noted upon the small-scale map of the Sierra channels at the conclusion of this report that in many places these diverging and converging channels appear to run uphill. The fact that these channels are originally of glacial origin, explains very clearly what can not be explained by assuming a pure fluviatile origin. This is the difficulty which can not be surmounted by advocates of the pure stream bed theory of origin of these channels. A notable case in point is the system of channels at Little Grass Valley in the La Porte district. From the McFarland Mine above La Porte it is noted that the shingling of the gravel appears to have two directions from one high point in the bedrock. In connection with this district, a sketch map, made by Mr. Stretch, an old-time engineer and surveyor who resided in the district for many years, is submitted. See Plate II. This sketch, to

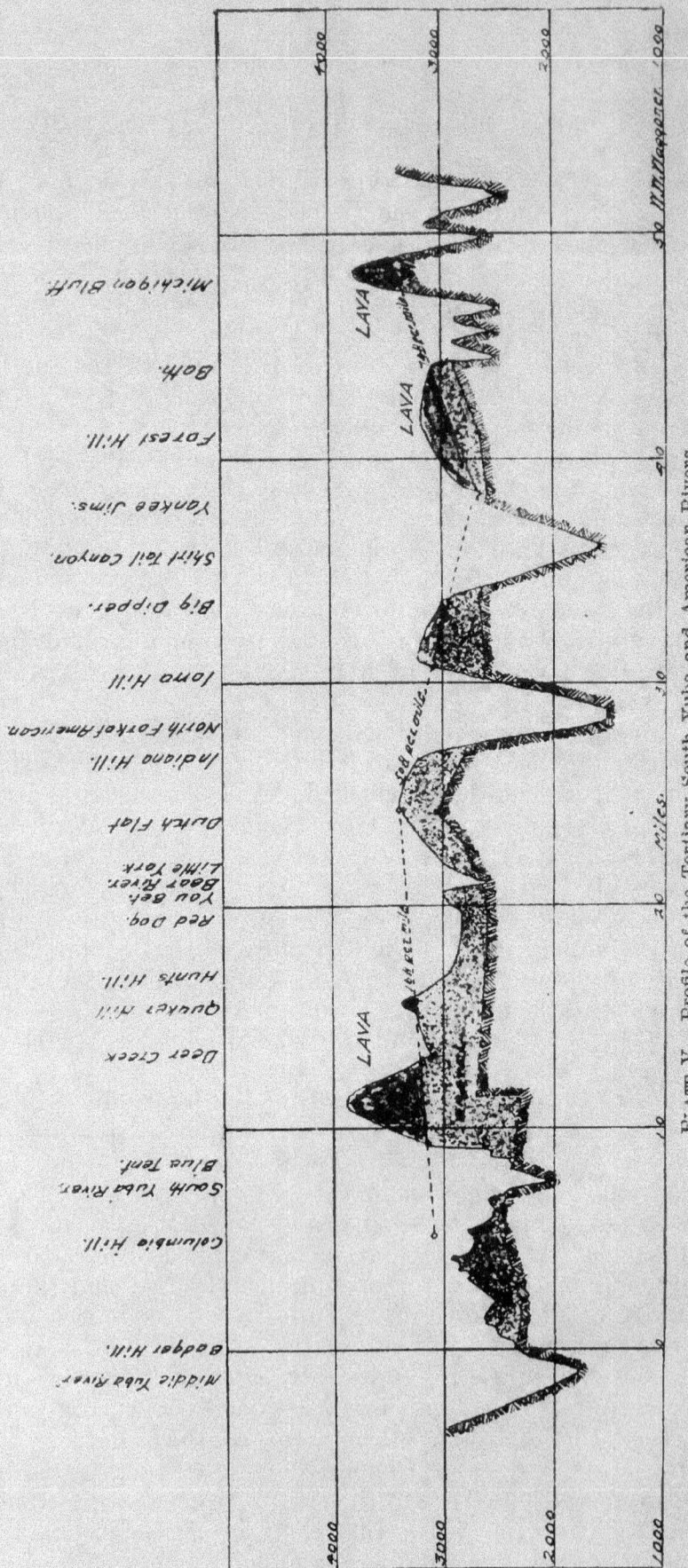

PLATE V. Profile of the Tertiary South Yuba and American Rivers.

the writer's mind, contains more accurate information with regard to the courses of these channels than is obtainable from any other source, and proves almost conclusively the original glacial origin of the channels in the higher Sierras, together with subsequent stream distribution of the gravels.

The region discussed in this section is probably the most important from an economic standpoint of any of the placer mining ground remaining in the state. The water rights controlling this section and the ditches governing it are largely in the hands of private power corporations and public service utilities. For this reason it might possibly be thought that the operation of the hydraulic mining in this region would be opposed to the interests of the big power companies. As a matter of fact this is not the case for the reason that not one-tenth of the water available during the hydraulic mining season is now being used by the power and irrigation companies. During the months from December until June the excess water going down the Yuba and the American rivers is several times the amount which would be required to wash all the available gravel in this region, large as that amount may seem.

The increasing of water storage facilities, the building of the impounding dams necessary for holding not only the miner's tailings but the natural erosion, which far exceeds what the miner develops, will furnish to both the power and the irrigation companies water and power far in excess of any of their needs for many years to come, and the mutual benefit derived therefrom will result in the complete harmony of all apparently conflicting interests.

During Cretaceous times the equivalent of the American River was undoubtedly much smaller than the present stream, as the greater portion of the area now drained by the north and middle forks was then drained by the great south fork of the Cretaceous Yuba. The lower drainage of the American is now expressed in channels of two types; one lying deep under the Quaternary alluvials of the present Sacramento Valley between Roseville and Auburn, and the second or intervolcanic phase lying along Boulder Ridge. Both of these, by their erosion, have created a certain enrichment among the Quaternary gravels, which is of some economic importance. In the neighborhood of Loomis there is a very considerable body of gravel which will run from 20 to 30 cents a yard, in which the values have evidently been concentrated by the erosion of the original shore gravels. This concentration was of irregular depth, in most places, being quite shallow, and covers several thousand acres in the aggregate. It might possibly be worked by an adaptation of the hydraulic method, with centrifugal pump stepped into the line to furnish pressure. It is possible also that some of this gravel might be worked by the use of steam shovels or other mechanical handling.

The original Cretaceous channel of the American River probably had its outlet somewhere near what is now Secret Ravine just east of Roseville. Going up stream in a northeasterly direction we find traces of the channel southwest of Loomis and at Rattlesnake Bridge. From here the channel runs almost due north to Auburn, where it makes a sharp bend southward again across Knickerbocker Creek to Pilot Hill. The Tertiary phase of this channel is expressed on Boulder Ridge and the ridge to the north of Dutch Ravine. Some concentration from this sluice robber channel was responsible for the placer diggings at Gold Hill and at Ophir, although local stringer enrichment was also responsible for a large part of the values.

In the vicinity of Auburn some possible drift ground still remains in the old channel where it has not been cut away by the intervolcanic channel, and lower down near Loomis are several acres of good hydraulic elevator ground, which is fed from this channel. From Pilot Hill in a southwesterly direction traces of this channel can be found near Clark Mountain and at Thompson Hill and Granite Hill southerly from Coloma. It does not, however, find its fullest expression again until it reaches the Placerville district. In this neighborhood there has been a great deal of enrichment due to the erosion of the original Cretaceous channel.

The Placerville district, situated on the ridge between the south fork of the American River and Webber Creek, has been a very rich field. The modern drainage, enriched by the eroded gravels, probably yielded the larger proportion of the gold which has been taken out, but considerable money was derived from drifting. It is estimated that about $25,000,000 has been produced from this area. A very full description of this area is given in Professional Paper 73 of the U. S. Geological Survey, to which the reader is referred for further information.

The original channel, which fed the Placerville Basin, has never been bottomed from Texas Hill west, and this could only be done by a long tunnel on the Webber Creek side. Webber Creek itself offers some dredging possibilities due to the old tailings which have been reconcentrated therein from the washings of the Placerville district. Dredging operations, however, might possibly be hampered by the narrowness of the Webber Creek channel at irregular intervals. In the neighborhood of White Rock Canyon, immediately north of Placerville, and in Randolph Canyon, is still some good hydraulic ground. In the former section mention was made of two tributary channels which came in from the north and were mostly eroded by the drainage of Rock Creek and of Silver Creek. Undoubtedly the channel which came down from Kentucky Flat contributed very greatly to the enrichment of the Placerville district, as it crossed the pocket belt which extends in a general way from the Georgetown district down toward Placerville. The district immediately north and south of Georgetown has produced a great deal from pocket mines, and the erosion of gravels of modern streams still contain a very considerable gold content. From Placerville east to Newtown this channel crossed the north fork and the south fork of Webber Creek. At Newtown considerable hydraulic mining was done. A branch apparently came into this channel from Pleasant Valley, which contains a very large quantity of gravel, in portions of which dredgeable areas could undoubtedly be segregated. This is especially true of the region around Newtown and immediately to the south. From Newtown on, this channel passed in a northwesterly direction on the ridge between the south and north forks of Webber Creek, and may be picked up at Clayton and again near Pacific House, where it has been drifted more or less unsuccessfully. It was joined from the north by the Silver Creek tributary, both forks of which had their origin on Pilot Creek above and below Forni's Ranch. From Pacific

House a channel can be traced in the canyon of the South Fork of the American River on either side almost up to Ralston Peak, but is of little economic importance. A tributary from the south joins this channel near Bullion Bend, which probably had its origin near Round Top and came down through Hell's Delight Valley.

From an economic standpoint by far the most important portion of this channel lies in the neighborhood of the Placerville Basin. Apparently there still remains a possibility of drifting in this district, which might be profitable if drainage could be gained at the bottom of the Deep Blue lead. To the northwest of Placerville the whole of this stretch has been drifted except for about a quarter of a mile south of the Gas Pipe claim. According to Lindgren, from Smith Flat to Alta near the Landaker tunnel, the channel's general course is from east northeast to west southwest for about two miles. The grade from White Rock to Prospect Flat is 39 feet to the mile; from Prospect Flat to the Linden Mine the grade is 53 feet to the mile, but from Texas Hill on southwesterly the channel has never been bottomed and the grades are unknown.

As there is very little of value from an economic standpoint on this channel, it is hardly worth while to go into much more detail regarding it. The reader is referred again to Lindgren and to Whitney's original report on the auriferous gravels of California for information which may have a historic value, if not an economic one.

The large dredging field in the delta gravels around Folsom has been described in detail many times. The reader is again referred to the bibliography.

SECTION 4.

COSUMNES, MOKELUMNE, CALAVERAS AND STANISLAUS RIVERS.

Along the old shore line, considerably north and even a little south of Ione, is very much delta gravel which has been deposited by the ancient river in its migratory discharge. Where modern gulch erosion concentrated these shore gravels a very considerable proportion of it was worked in a small way. It might still be possible to handle some of these shore gravels at a profit.

The lower reaches of this river are almost entirely eroded above Ione, although the junction of its north and south forks left a little gravel at Irish Hill. Above this, on both forks, the channel is nearly always, where present, heavily lava capped except on the upper reaches of the north fork, and also near Volcano on the south.

At Oleta the north fork is concentrated into the gulch made by modern erosion, and very good placer diggings resulted. Below Oleta the channel is found in place. Following up stream by way of Aukum it passes west of Cedar Creek to Fair Play. Above Fair Play modern erosion again produced a concentration in the gulches on both sides of the main river. At Omo Ranch the channel is again lava capped, also at Slug Gulch. There is considerable virgin drift ground here which has some possibilities; in many places, however, the gravel has been so completely cut out by the lava flows that, even including bedrock pay, there is not enough in the channel to justify working. At Menden and at Indian diggings are considerable bodies of hydraulic ground, but there is not a great deal of water for working purposes. A branch stream came in from Grizzly Flat by way of the Henry diggings. This stream still has some good drift ground on it. A lower course of this same river goes down from Menden through Boughman's Mill. Again joining the main channel, it passes out through Coyoteville.

Following up the south fork of the ancient river from Irish Hill its course is very nearly the same as that of Dry Creek. At Posey Hill is a drift channel which has considerable possibilities. At this point the main channel is joined by another branch which comes down from the northeast. Still another channel comes through Volcano, where it has been largely eroded into the gulches to join this main channel by way of the south side of Dry Creek. Passing upward through Lockwood, where it is joined by still another tributary, it continues easterly for several miles along the ridge. From Posey Hill on up, the channel is heavily lava capped and apparently the greater portion of it is virgin. There is not, however, a great deal of water to work with and it is extremely doubtful if this channel can be called drift ground except in certain localities.

Coming down to the extreme lower reaches of the ancient river, which have been widely distributed over the rolling hills of the Sacramento Valley by modern erosion, on the Plymouth Road between Michigan Bar and Forest Home, there is still a great yardage of gravel which could be hydraulicked, and also several thousand acres of shallow ground in which it is possible that portions might be segregated which will prove economically profitable for a small dredge. Only a portion of this ground is deep, as in most places the bedrock comes close to the surface. The only possible water for working this ground would have to be obtained from the Cosumnes River. This ground is within 25 to 30 miles from Sacramento by wagon road.

As stated before, the exact course of this channel is rather hard to determine in its lower reaches. Many of the creeks above Ione, including Horse Creek and Mule Creek, were worked in the early days and paid very well. They were undoubtedly enriched by the erosion of this channel. From Plymouth on up to Volcano there have been several fairly good hydraulic diggings, notably on the headwaters of Rancheria and Dry Creek. At Volcano was probably the greatest enrichment. Here the bedrock is largely limestone, and the gravel and lava were removed by erosion, concentrating the values in the channel to such an extent that they formed what were at one time some of the richest placer diggings in the State of California. Above Volcano to the north, the Elephant Mine has been operated within the last few years. Considerable gravel still remains in these diggings. From Volcano on up to the head of Ashland Creek, the channel has not proved of any great economic value. A branch running down to the east of Pine Grove has been hydraulicked to a certain extent. Most of the bedrock here is in the Paleozoic metamorphics, but there has not been a great deal of local or stringer enrichment.

The north fork of the river, passing up through Bridgeport, has been worked at various places and, between Fair Play and Grizzly Flat, it crosses a pocket belt which has been responsible for considerable enrichment. This channel can be traced up the ridge between Long Canyon and Steeley Fork for a considerable distance, and some mining has been done on its upper end.

The south fork of this channel extends much farther back into the Sierras. From Lockwood up to Hams its course can be traced by a heavy lava cap. Slightly above Hams it divides, one branch going toward Leek Spring Hill and the other running almost due east toward Mokelumne Peak. Neither of these branches are of any economic importance.

The Cretaceous equivalent of the Mokelumne River is probably the most important of the southern rivers. Starting with the Concentrator channel near Amador City, which was extensively worked by hydraulic mining in the early days, the ground has been drifted down through Jackson and Butte City to Mokelumne Hill. Below this point it is joined by a number of channels which are partly virgin. All of these converge at Central Hill. In Chili Gulch, below Mokelumne Hill, there is still some good hydraulic ground, and also to the west of this. The main channel, which passes down near Fosteria, is practically virgin. The gravel is very deep, with intermittent lava capping.

The Tunnel Ridge channel, the Duryea channel, and the Blue Lead still have considerable virgin ground which might possibly pay for drifting, but in which a good deal of water would have to be handled. Water for mining purposes would probably have to be obtained from the Mokelumne. At Central Hill these last named channels unite with the main branch which comes across from Altaville, through San Andreas. Near Calaveritas this channel is joined by the Fort Mountain channel. Starting on the west side of Tiger Creek, and going through on the east side of Bald Mountain there is considerable hydraulic ground on this channel. On the other branch, coming down east of West Point through Railroad Flat and Fort Mountain and down to Sheep Ranch, are several miles of virgin drift ground. At Rigneys, near Sheep Ranch, and below, is a considerable body of hydraulic ground of a grade and feasibility of working that might make it attractive to the small placer miner. From Sheep Ranch on to Calaveritas, erosion has removed a good deal of this channel, although at Cave City and at Old Gulch considerable hydraulic mining has been done. Coming back to the main south fork of the ancient river we follow it from Fourth Crossing through Dogtown and Angels Camp. Considerable drifting and hydraulicking has been done in this neighborhood. The headwaters of this branch are on the north and south sides of the north fork of the Stanislaus River. It is again picked up about six or eight miles above Avery and follows down the ridge in a southwesterly direction to Douglas Flat. From here it is joined by a very rich tributary which comes down from above Murphy's Ranch. At Vallecito it is joined by a northward flowing tributary which came through by way of Columbia. At Vallecito there has been a great deal of erosion and some exceedingly rich placer diggings have been developed.

At San Andreas is still untouched drift ground of possible value. Below Central Hill the channel swings into the old shore line near Valley Springs. It has been reconcentrated in the Calaveras River, on the lower reaches of which considerable dredging and placer mining has been done. Continuing on westward to the region of Campo Seco, Comanche, and Lancha Plana, the delta of this river was exceedingly rich, and was much worked in the early days. The low, rolling hills in this region are covered with shallow gravel, which will run anywhere from 20 to 50 cents a yard. There is an excellent opportunity here for either hydraulicking with centrifugal pumps or for working with a steam shovel plant and belt conveyors. The average richness of this ground and the uniformity of the gold distribution make it attractive, and it will undoubtedly be worked at some future date. Some dredging has been done near Campo Seco.

On this channel of the ancient Mokelumne are probably upwards of 400,000,000 yards of gravel, a large portion of which has a good chance of being worked at a profit by either dredging, drifting, hydraulicking or mechanical methods, such as steam shovel work.

Lower down, near Jenny Lind, a southward flowing delta of this same river has concentrated much fine gold, and a good dredging area has been developed here, which is now almost worked out. From Jenny Lind north along the shore line of the ancient ocean, there is still a good possibility of finding shallow areas of ground which might be worked either by mechanical means or by hydraulicking with reservoirs and centrifugal pumps.

Coming back to the south fork of the South Fork of this ancient Mokelumne River, we have an area between Vallecito and Yankee Hill which has been one of the richest from the point of production in the State of California. The bedrock here is largely limestone, and the channel was concentrated in many places to the point of exceeding richness. In the neighborhood of Columbia, alone, over $55,000,000 has been taken out. The course of this channel can be followed up stream across Woods Creek, where it makes a sharp bend to the north through Yankee Hill and again crosses the south fork of the Stanislaus River. The last trace of this branch is found near American Camp where it was hydraulicked. The Stanislaus River was tremendously enriched from about the junction of Five Mile Creek on the south fork clear down to Melones, by the erosion of the branches of this channel. After crossing Woods Creek, the main branch of this channel can be followed almost due easterly through Phoenix Lake by way of Browns Flat to Arastraville, where it makes a sharp turn to the northeast, and can be followed up the ridge between the north fork of the Tuolumne River and the south fork of the Stanislaus by way of Confidence, Sugar Pine and Long Barn clear up to Cold Spring. This channel has never been of economic value from the drifting standpoint, though considerable work has been done upon it.

At Melones, near the junction of Coyote Creek, is considerable local enrichment, due to the crossing of the pocket belt, which comes through by Morgan Hill. This, in addition to the gold which was brought down by the Stanislaus River from the crossing of the old channels four or five miles above, resulted in some very fair surface placers.

In connection with this river, it is advisable to trace the course of the channel which has probably been one of the most effective sluice robbers and the cause of more blasted hopes among placer miners than any channel in the State of California. This is what is known as the Table Mountain channel. Starting in on the ridge below Clover Meadow, it runs southerly across on the Middle Fork of the Stanislaus to Shotgun. From here it turns westerly by way of Mount Knight and Collierville toward Douglas Flat. This channel was a strictly intervolcanic channel of late Neocene age. For this reason it contains no values until it meets the old channel below Douglas City. Crossing down slightly to the east of Vallecito, it follows the bed of the old Columbia channel for several miles. As this was an exceedingly rich ancient channel flowing northward, and the Table Mountain channel flowed southward with its course cut out and carved by frequent rushes of volcanic lava, the channel was enriched to the extent of its ability to rob from the old Columbia channel for a very considerable distance beyond the point of its departure, which occurred somewhere in the neighborhood of Parrott's Ferry on the main Stanislaus River. Continuing southward west of Springfield and Shaws Flat, this channel can readily be traced under what is now known as Table Mountain through the ridge west of Jamestown. Wherever it has been eroded by modern drainage there has been a certain amount of local enrichment, but practically all of this enrichment was undoubtedly derived originally from the old north-

Photo No. 33. Near Soulsbyville, Tuolumne County.

ward flowing Columbia channel. At Montezuma and Mountain Pass this channel has been worked by hydraulic mining and has also been drifted. At numerous places north of this, notably at Springfield, much money has been spent in an attempt to drift this channel but none of the attempts have been successful.

A characteristic of this channel is that numerous parallel courses of shallow depth have been cut out and then the whole covered by an enormous flow of latite. For this reason prospecting it has proved tremendously expensive. The course of this channel can be traced by this latite flow from Montezuma down through Peoria Mountain, on the southeast side of the present Stanislaus River, clear down to Knights Ferry, where it probably emptied into the old Tertiary ocean somewhere near Wildcat Creek. At various points along this line it has been mined, notably at the Wagner Ranch, Peoria Basin and on Owl Creek, but none of the operations have ever been successful. Judging from its history, this channel should be avoided from a mining standpoint.

Coming back to the Columbia channel, in its course up toward Yankee Hill, is still some possible hydraulic ground; also on the main Stanislaus River at the junction of the south fork there is a deep hole which might possibly pay to prospect with a view of turning the main river during the summer time and working the gravel. It would probably be a very expensive operation, but there is undoubtedly considerable gold in the main stream at this point, due to the erosion of the rich channel which crossed above it on the south fork.

No recommendations have been made with regard to debris dams on either of these streams. Owing to the rolling nature of the country, it seems that more likely dump sites could be found in the uplands above the streams, and if hydraulic mining were resumed, it is probable that very satisfactory brush and crib dams could be constructed without the necessity of the expensive concrete construction which is absolutely mandatory in the northern rivers. As the water rights in this country are not nearly so good as the water rights to the north, it does not seem very likely that hydraulic operations would ever assume the proportions, even relatively, in this region that it would in the region of the American, the Bear and the Yuba. On the other hand, there is no doubt that there still remains in this region several hundred million yards of gravel which are adaptable to handling by hydraulicking, or by the use of some form of mechanical elevation. In addition to this, there is still considerable virgin drift ground—notably in the region around Vallecito, San Andreas and Mokelumne Hill.

TUOLUMNE AND MERCED RIVERS.

The Tuolumne River is the last of the great rivers toward the south which drained the western slope of the Sierras during Cretaceous and Tertiary times. A shore outlet of this river was apparently located at what is now Chinese Camp. At this point the gravels were distributed and reconcentrated by modern erosion, and were exceedingly rich. Although little water was available for working them, an entire hill of gravel was broken down and hauled away, operations being conducted in the simplest and crudest manner, but with considerable profit to the owners due to the extraordinary richness of the gravel. Apparently another outlet of this stream came through to the southeast of Chinese Camp by way of what is known as the Mencke Ranch. To the north of this considerable handwork was done on the concentration of the ancient gravels of the river. There is still a possibility of a drift mine in this locality, although the amount of water in the course of the old channel makes it rather expensive working. Following up stream the ancient channel goes back almost along the present cañon of the Tuolumne River. At various points are still segments of uneroded gravel, notably on Big Humbug Creek, and a trifle to the north of Smiths Station.

To the north of Groveland, on the head of Big Humbug Creek, is an area of exceedingly spotty gravel, from which the lava cap has largely been eroded. This has been worked with very little success for several years. From here on up to the Gravel Range, on the east side of the present Tuolumne River, are several deposits of the gravel of the ancient stream, which apparently have not been considered sufficiently profitable to do much work on. As in this neighborhood we are going away from the main belt of metamorphic rocks and approaching the granitic area, this may possibly be the reason that the value of the gravels appears to have greatly decreased.

Above Jacksonville on Moccasin Creek a great deal of work was done by the early miners on the light gravels which were concentrated in that creek. It is quite possible that another course of the Tuolumne River, which has now been completely eroded, may have contributed toward the enrichment of Moccasin Creek.

At La Grange, on the present Tuolumne River, and in the immediate vicinity, there is a tremendous amount of delta gravels which were apparently fairly profitable during the days of unrestricted hydraulic mining. Many mines were operating here during the 70's and 80's, and at present along the bed of the present river the ground is being dredged at a fair profit. From one to two miles above La Grange there are still large banks of gravel which are capable of being hydraulicked. The gravel is slightly cemented and not very heavy. It is possible that this area, instead of being a delta of the ancient Tuolumne River, may have been built up by a minor stream which, more or less, corresponded to the present Merced River in its drainage. As the country has been heavily eroded to the eastward, it is exceedingly hard to identify this delta deposit with any of the ancient streams.

On the Merced River there has been some very fair gravel at Snelling, where a dredge was operating for several years. Between Snelling and Merced Falls, in the early days, considerable surface placering was done. From here on up to Horseshoe Bend the river had considerable grade, but a small amount of placer work was done in Pleasant Valley. At Horseshoe Bend efforts are still being made to mine the stream. Above Bagby and in the tributary gulches in that neighborhood some hand mining was done in the early days.

There are traces of an ancient channel above Coulterville, about four or five miles south from Mountain King and on the southeast side of Mount Bullion. Whether this channel was continuous through these areas, or whether they are fragments of several small channels, is a matter that is impossible to determine, due to the extensive erosion which has taken place. There is no question, however, that most of the modern enrichment of the present Merced River has been derived from the pocket belt which extends northward from Mariposa toward Coulterville at Bear Valley. A few miles south of Bagby extensive placer work was done in the early days, but the richness of the gulches in this neighborhood was undoubtedly caused by the primary concentration of quartz seams and stringers bearing fairly rich pockets in this neighborhood. The same thing applies to the placer work on Hornitos Creek and Burns Creek in the neighborhood of Hornitos. This placer was fairly good in the gulch diggings in the early days, but it has now been pretty well worked out. The Pleasant Valley gravel is said to run better than 30 cents a yard. If this be true, it is strange that it has not been worked, as there is a considerable yardage of available hydraulic gravel still to be seen. Above Coulterville at Dogtown hydraulicking is now being used to sluice off an area of seams and pockets which has made placer diggings by its erosion. From Bear Valley down to Mariposa are local concentrations in the gulches caused by the erosion of pocket veins on the Mariposa drainage. These gulch diggings have been worked by hand, and it is quite possible that some of them would still pay to hydraulic.

In the main, the Merced region consists of very spotty gulch diggings, which are mostly worked out. The channel mentioned above, which passes about five miles east of Coulterville, has been slightly drifted, but without much success: for one reason because the cuts which run into it did not reach the bottom of the old channel.

UPPER SAN JOAQUIN, FRESNO, KINGS, KAWEAH AND KERN RIVERS.

On the upper San Joaquin, which forms the boundary line between Madera and Fresno counties, there has been some slight enrichment which was worked in the early days, notably at Italian Bar and in the neighborhood of Huntington Lake and Cascade. There have been some placer operations, but no great amount of money has been taken out.

In Madera County, on tributaries of the Fresno River, considerable placer work was done in the early days. Grub Gulch was worked for several miles toward the river, but it is now all worked out. The same applies to Coarse Gold Creek. On Kings Gulch, above Grub Gulch, is a fragment of an old channel which has been unsuccessfully drifted. This may possibly be connected with the segments to the north which have been noted on the Merced River, but probably is independent. At Friant, on the Fresno River, the gravel runs a few cents a yard, and the saving of from one to two cents a yard is now being made as a by-product in a gravel crushing plant located there.

On the Kings River, about twenty miles above Piedra, is a gravel deposit on which considerable money has been spent in a futile manner. This deposit drains the granite belt almost exclusively and is exceedingly low-grade. The front rims, however, in the early days paid small wages to miners where there was a certain amount of local concentration, and as a result of exaggerated stories regarding the richness of these diggings, numerous companies have been floated without the slightest prospect of success.

On the Kaweah River, in the Marble Fork, attempts have been made at placer mining, but none of these have been successful. Practically all of the gold in the gravel, which is extremely low-grade, has come from the breaking down of pyritic masses from the granite, which is the predominant country rock.

Around Woody, in Kern County, and White River, in Fresno, the erosion of stringers in the granite has produced local gulch concentrations which were extensively worked for considerable distances in the early days. Very little of this is left, however, which can be classed as placer ground. There have been some good quartz ledges in this district, which were probably responsible for the feeding of the placers.

One of the earliest placer camps in the state, at Keyesville on the upper Kern River, consisted mainly of gulch diggings, which were enriched by the erosion of stringers in the granite close to an area of metamorphic rocks. In Rich Gulch, Sand Gulch and Keyes Gulch there is evidence of much work, but little or no workable ground is left. For thirty miles down the river, however, the gulches have been worked at intervals. Around Havilah is much gravel, but the pay has already been worked out. On Piute and Greenhorn mountains is still a little gravel which might possibly pay to work.

On the south fork of the Kern River, about thirty miles above Isabella, there is a little hydraulic ground with poor dump, which is not of very high grade. Attempts, however, have been made to work it. The river sands below the junction of the north and south forks of the Kern River carry some telluride combinations of gold. There is a gorge in the river below Isabella and commencing just above Keyesville, which could easily be worked by diverting the river. This river is reported to be virgin at this point on the authority of the men who tried to work it about fifteen years ago. It seems quite possible that this gorge might be worth while prospecting, as much gold must have been fed into this river from the gulch diggings on the northeast side. It would not be a very expensive undertaking to prospect it.

SECTION 5.

OUTLYING DISTRICTS.

While the Coast Range in the Franciscan rocks has never been productive of any great concentration of gold, nevertheless there have been numerous local enrichments in which extraction has been attempted at various times by placer miners. In San Luis Obispo County is an area of Franciscan metamorphics which has thrown some gold into the tributaries of Poso Creek, but not in any great quantity. Attempts have been made to work this by steam shovel and by mechanical elevation. At Fraser Canyon and near La Panza are considerable areas of gravel but they are low-grade and extremely spotty. Thus far no economic success has attended any work done in this district in spite of the fact that there is plenty of water available for working. The gold is exceedingly flaky and fine and the district is very spotty.

West of Jolon, in Monterey County, is a pocket district close to the coast which, by its erosion, has produced some very heavy gold. Some of the finest nuggets that the writer has encountered in the whole state have been exhibited to him as coming from this region. Apparently, however, there is no considerable amount of placer territory which would justify the expenditure of any great capital or an attempted working of this district by anything but pick and shovel methods.

At Surf, Santa Barbara County, is a very considerable concentration of black sand which is slightly auriferous but which is not an economic proposition.

Returning to the northern portion of the Coast Range, in Mendocino and Lake counties, there are local concentrations similar to those already mentioned at Poso and La Panza. While in the northern country these areas carry some platinum, their gold content is so low that they are not of any economic importance. There is one deposit in the neighborhood of Hopland which carried considerable platinum, but the distribution of both gold and platinum was so spotty and irregular that an attempt made to work it by modern mechanical methods failed of success.

There is only one of these outlying districts away from the great belt of metamorphic rocks in the Sierras which seems to the writer to have any economic interest. In Mono County, on the headwaters of the Walker and on Virginia and Dog creeks, is an area of gravel which will pan about 25 cents a yard. At Bodie Flat, not far from this region, considerable placer work was done in the early days. While the Bodie concentration appears to have been principally a primary enrichment from ledges and stringers, the area on Virginia and Dog creeks consists of well-rounded wash gravel, varying in depth from five to ten feet and distributed over several hundred acres of ground.

At the time of the writer's visit the depth of the gravel did not appear to be sufficiently great to justify the expense of a hydraulic installation, owing to its widespread distribution, and the difficulty of getting the water upon the ground under pressure. Since that time, however, it is reported that shafts have been sunk through what appeared to be the bedrock of the country, proving it to be false bedrock. It is said that about thirty feet of gravel has been developed underneath this false bedrock, which carries fairly good values. If so, the district is worthy of investigation, with an eye to the possibility of working it with a small dredge, as the upper gravels apparently carry gold in sufficient quantity to make them of economic consideration.

DRY PLACERS.

Dry placers of the State of California are for the most part located in the southern portion, chiefly in the region of the Mojave and Colorado deserts. These were probably the earliest known sources of gold in the state. In connection with the dry placers of this region, there is one district, however, that has not heretofore been discussed but, as it lies in the same general region, it will be briefly described.

This district forms what is probably the only promising or possible area of ground in southern California which is suitable for dredging. It lies in the Holcomb Valley about four miles north of Big Bear Lake. It is apparently a primary concentration from the erosion of the schists and porphyrys of the Gold Mountain region. It is a broad porphyry belt which crosses the country in a northwesterly and southeasterly direction and can be traced for several miles. This belt is apparently full of numerous quartz stringers and pocket seams. In places the country rock for considerable widths will run three or four dollars a ton in gold. It is from the erosion of this belt that the Holcomb Valley placers have been formed. During the late 80's and early 90's it was operated by an English company by means of steam shovels and elevators. Most of the work was done around the edges of the deposit and but little was attempted in the deeper gravel in the center. However, it is said that most of the material handled by this operation averaged between 30 and 35 cents a yard. This seems quite possible from the nature and type of the erosion.

At the present time there is an area of some three or four hundred acres on Holcomb Creek, which varies in depth from ten to fifty or sixty feet, and which seems as if it might be possible to segregate sufficient dredging ground to make an economically feasible proposition. The only water available is that from Holcomb Creek, but a minimum of thirty to forty inches is at all times available and during the winter and spring months there is considerably more. This water should be sufficient to maintain the dredge pond during the greater portion of the year. The proposition apparently justifies prospecting to determine its possibilities.

Coming back to the dry placers proper, a good deal of prospecting and work has been done in the San Gabriel Canyon and, in the late 80's, several small operations were in process on the auriferous gravels that flank the mountain sides. These are undoubtedly of recent origin and of primary concentration. The amount of water available for working them was very limited, but for the most part they were worked in the spring when impounding of water was possible.

From the time of the days of the Mexican colonies gold has been won from the dry placers of the desert. The Indians in the days of the padres used to bring in gold to the missions, which was laboriously recovered from the desert washes by the crudest of methods. Unfortunately, however, there is no reliable record of production until comparatively modern times; but we know that in total it amounted to many millions of dollars. Since the American occupation of California, we have, however, somewhat of a record, and we know of districts which have produced from a hundred thousand to two millions of dollars—all won by hand methods and mostly by dry washing.

It should be remembered, however, in consideration of these figures, that the values won from the desert placers have generally been earned by the operations of a multitude of men working independently and that the extent of their operations in point of time covered a great many years; for instance, in one district, the Potholes district near Yuma, from which a reported production of $2,000,000 has been taken out, as many as 400 or 500 Mexicans were working with hand washing machines for a period of several years. If this point is borne in mind in considering the gold production of the desert, it will be readily understood why it is that to date not a single large-scale operation handling dry placers in southern California has ever been successful.

The results of the writer's investigation, which has covered every dry placer district of importance in the desert, have on the whole been decidedly disappointing, and the writer is forced to the conclusion that these placers have already been exploited by the only practical means: the Mexican with his little hand-operated bellows and rocker. The reasons for this conclusion will be given herewith. From the standpoint of geological history, the district south of the Tehachapi has been distinctly different from the region of the great valley to the north. The ranges of subsidence and elevation have been much less during Tertiary and later times. During the later ocean transgressions, including the Ione, which covered the foothills of the western slope of the Sierras and of the Coast Range, this region was comparatively quiescent. For this reason it may be assumed that the topography and to some extent the climatic conditions of this region, during the time in which changes were taking place to the north, did not vary greatly.

This fact leads to a vital difference in the manner and amount of deposition of the gold in the desert channels, so far as it occurs along the line of channels. Whereas in the north we have a primary concentration from the erosion of the metamorphics and intrusives of the Sierras, to be followed later by a reconcentration of the values in the Cretaceous and Tertiary streams by the present, or Quaternary, drainage running in many cases normal to the older, in the region of the desert we have an entirely different condition, and one not nearly so favorable to the concentration of gold in large quantity and persistent amount.

In the south, the drainage during Tertiary times evidently did not greatly vary from the present lines. Roughly, practically all of the deposits on which stream action had the slightest bearing may be classed along the drainage lines of ancient streams whose courses follow the general direction of the modern ones, such as they are. In some cases, the ancient ones seem larger than those of the present day—in other cases smaller.

In San Diego County, we have a course which, commencing in the Cuyamaca Mountains, near Julian and the Banner district, runs down through Ballena via Hatfield and Coleman creeks—which have crossed it and caused local concentration—down through Ramona and the San Vicente region, and finally empties into the ocean somewhere around La Jolla. This region is not of economic importance, as the values were very spotty, according to the records of the districts, and facilities for working are limited to a very short season of the year.

East of this, in Imperial County, we have either local erosion deposits, like that in the Borego country—which is said to average about fifteen cents a yard—or the deposits in the Chuckawalla basin, the Chocolate Mountains, the Eagle Range, and the country north and east of Desert Center. These were all more or less tributary to the drainage system of the Colorado, which is probably one of the oldest rivers on the western slope. On the Colorado itself, or near it, we have the Potholes, at Laguna; the Picacho Basin; and other deposits in the Blythe-Parker-Ehrenberg region. These were all investigated by the writer; and his conclusion is that, although a recorded production running into the millions has come out of them, little remains of interest to either large capital or the small miner.

The gravel, or eroded material, was very spotty and seemed to carry the values in the subangular stuff at some distance above bedrock; and any pumping scheme which might be used to get water would be so costly and cumbersome that the ground will never pay it back. The Mexican with his little portable machine and lack of overhead in moving from gulch to gulch where the greatest concentrations were, had the only feasible system. The Picacho and Laguna deposits are disintegrated schists and slates for the most part, containing quartz stringers from which the gold has come. In both cases the erosion caused by the Colorado River at the base of a low range of hills has resulted in the accumulation of this material. The river wash itself does not carry much value, but the disintegrated matter above carries it; and the erosion of present day gulches has resulted in local concentrations which were worked by the Mexicans.

In the Owens Lake country, along the eastern base of the Sierras, is a series of short channels and delta gravels along the shore line of the ancient sea, which once occupied the Owens River Basin. Apparently a short channel came in through Red Rock Canyon, where it was joined by another which came from the west in the neighborhood of Tehachapi by way of Jawbone Canyon. The shore line can then be traced clear through by way of Goler and Summit Diggings to Copper Canyon and Long Range. It is especially developed at Coolgardie, but is again lost as the formations dip under the recent wash near Barstow. At intervals, notably at Goler and at Coolgardie, there appear to have been short channels whose deltas are expressed in these places.

In addition to this, there are washes of recent origin, such as the one at St. Elmo near Goler. The St. Elmo wash was directly enriched in modern times from the erosion from the Stringer district, and is still being worked in a small way by hand dry washers. Another short channel comes down from the Goler wash in the Panamint Mountains.

As this wash is typical, a photograph of it is shown. This system of channels and of shore gravels is one of the best known in the south and corresponds, in the nature of its formation, to the auriferous and conglomerate gravels which have been noted in Siskiyou and Shasta counties. The conditions of its formation were, however, widely different. This system illustrates a point which strikes the writer very forcibly in its bearing on the possible economic operation of these gravels.

As a rule, the stream gravels, which consist of well-rounded boulders of crystalline rocks from the Sierras, carry no value at all. Wherever

PHOTO No. 34.   Head of Goler Wash, Panamint Mountains.

they cross a belt of metamorphics containing quartz stringers, or any body of olivine bearing, or other basic rocks, by local erosion there is produced, in connection with the heavy cloudbursts that sweep down the gulches intermittently, local concentration of gold. As a result, at Goler, for instance, one can notice that the shingling of the washed material is transverse to the general direction of the course of the deposits. Heavy gold is found near the tops of the ridges and from 30 to 60 feet above bedrock and on local false bedrocks of clay and disintegrated volcanics.

At Summit Diggings, intensively worked on account of its shallow depth during the early days, operations have been started which contemplate bringing water twelve miles to cover the deposit. The same thing has been proposed at Coolgardie with a fifteen mile water system. As the richer portion of the concentrations, that is, the portions which have accumulated in the present day gulches, have long been worked

Photo No. 35. Dry Washer (Close-up), Summit Diggings,
Kern County.

out by hand methods and the general body of the deposits in either of these places will only run from 15 to 60 cents a yard, it does not seem that either is an economic proposition.

In Lytel Creek, along the line of the Santa Fe, there was a local concentration from the stringers in the granite. This has already been worked out. A little water was available here.

In the prospecting of any of these areas it should be borne in mind that the greater concentration of values will always be found where there has been the greatest erosion. This concentration will naturally lie fairly close to bedrock, but in estimating the average values for the yardage available care should be taken to prospect the entire surface of the available ground and to carefully note the deposits of the richer strata of material. This characteristic is especially evidenced in the placers above mentioned on Lytel Creek.

The last of these very interesting systems consists of a channel which, coming down from Placerita Canyon near Newhall, is another branch which headed above Bouquet Canyon, crossed into San Francisquito Canyon and came down somewhere below Piru, to be joined by another course which came from Lockwood by way of Piru Canyon. This was

PHOTO No. 36.   Dry Washer, Summit Diggings, Kern County, Cal.

much worked in the early history of Los Angeles and Ventura counties, but, as even the Chinamen gave it up as no longer affording them a living, it would not be advisable for a white man to try it. The values appeared to be concentrated on the points of the present hills and in the gulches. Quite a little water was available for working this system.

Attention must again be called to the fact that, in estimating the gross production of this, as well as other regions of the desert placers, a large part of the work which has been done there was done when wages averaged about $1 a day in times of industrial depression. Many men, making a bare living at best, preferred to be their own masters while doing it; and the aggregate of all their earnings presents an impressive figure. Considering the length of time involved and the number of men employed, it is no wonder that the aggregate amount, even from very poor diggings, was large.

One final point that has impressed the writer, and with particular force, is the total and absolute failure of air separation processes when applied on anything larger than a hand scale. Two monuments to this failure still stand in mournful state; one at Goler, and the other at Coolgardie. Limited capacity and imperfect separation are the main reasons. A machine which will work on a laboratory scale with material which has been dried on a hot plate certainly will not be a success when applied to the general run of desert material, especially after a shower has moistened the ground. The material simply balls up on the canvas tables which are generally used. Furthermore, machines of this character are generally so cumbrous that they lose the chief advantage of the hand machine, which is its mobility. As the hand machine is light and easily operated, it can be transported from one spot or concentration of gold to another as soon as the pay has been worked out.

In the accompanying illustrations the simplest form of hand machine for working dry placers is shown. These photographs were taken during the prospecting of the Summit Diggings before referred to.

At the present time, near Randsburg, a dredge operation has recently been started in which the principles involved in the Huelsdonk concentrator, described in the second chapter of this volume, are being used. There is not much question that the saving should be satisfactory with this type of machine, but to the writer's mind the question remaining to be proved is whether the values are in the ground, as the concentrations in this district are exceedingly treacherous and spotty.

Bibliography.

Iowa Shaft—Port Wine District. Mining and Scientific Press, February 6, 1875.

Gold Run District. Mining and Scientific Press, March 13, 1875.

Hydraulic Mining on the American and Bear Rivers. Mining and Scientific Press, beginning July 24, 1875, and ending February 19, 1876.

Wisconsin Hill. Mining and Scientific Press, May 6, 1876.

San Juan Ridge. Mining and Scientific Press, May 13, 1876.

Fairplay—Cosumnes River. Mining and Scientific Press, June 17, 1876.

Forest Hill—American River. Mining and Scientific Press, July 8, 1876.

Michigan Bluff. Mining and Scientific Press, July 15, 1876.

Yield of Gold Placers. Mining and Scientific Press, January 20, 1877.

Gravels of California. Mining and Scientific Press, September 20, 1879.

Projected Weaver Basin Tunnel. Mining and Scientific Press, February 7, 1880.

Theory of Auriferous Gravel Channels. Mining and Scientific Press, October 9, 1880.

Spring Valley Mine. Mining and Scientific Press, December 31, 1881.

Dry Placers. Mining and Scientific Press, April 20, 1895.

Black Sands of Pacific Coast. Mining and Scientific Press, May 16, 1896; December 11, 1894; December 17, 1894; May 15, 1897.

Mining in Northern California. Mining and Scientific Press, February 6, 1897.

Recollections of California Mining. Mining and Scientific Press, January 29, 1898.

Old Channel Placers. Mining and Scientific Press, April 13, 1901.

Dredging. Mining and Scientific Press, January 5, 1901; January 12, 1901; November 2 to December 14, 1901; April 16, 1904; April 15, 1905; April 22, 1905; April 29, 1905; August 19 to September 9, 1905.

Drift Mining at Placerville. Mining and Scientific Press, July 12, 1902.

Beach Mining with Surf Washer. Mining and Scientific Press, June 6, 1903.

Gibsonville District. Mining and Scientific Press, July 29, 1905.

Calaveras Channels. Mining and Scientific Press, September 9, 1905; September 16, 1905.

Hydraulicking at Cherokee. Mining and Scientific Press, September 8, 1906.

Dredging. Mining and Scientific Press, May 4, 1907; August 15, 1908; April 15, 1911; October 7, 1911; October 14, 1911; February 24, 1912; March 2, 1912; March 9, 1912; March 8, 1913; November 8, 1913; November 22, 1913; February 22, 1919; April 5, 1919.

Ancient Channels. Mining and Scientific Press, June 8, 1907.

La Grange Mine. Mining and Scientific Press, October 10, 1908.

Placers in Ventura County. Mining and Scientific Press, March 5, 1910.

Hydraulicking in Trinity County. Mining and Scientific Press, July 30, 1910.

Hydraulicking on the Klamath River. Mining and Scientfic Press, March 28, 1914.

Dry Placer at Goler. Engineering and Mining Journal, August 29, 1896.

Dredging. Engineering and Mining Journal, March 13, 1897; November 20, 1897; December 11, 1897; December 25, 1897; August 4, 1900; August 11, 1900; August 18, 1900; June 29, 1901; July 6, 1901; November 7, 1903; March 31, 1904; July 7, 1904; December 8, 1904; July 15, 1905; July 22, 1905.

Dry Placers. Engineering and Mining Journal, July 8, 1899; May 9, 1903; August 29, 1903.

Gold in Tailings—Trinity River. Engineering and Mining Journal, January 18, 1902.

Wing Dam on the Yuba River. Engineering and Mining Journal, October 11, 1902.

Cretaceous Conglomerate of the Siskiyou Island. Engineering and Mining Journal, October 31, 1903.

Hydraulicking in Humboldt County. Engineering and Mining Journal, February 23, 1905.

Origin of Placers. Engineering and Mining Journal, June 1, 1905; June 22, 1905; June 29, 1905.

Dredging. Engineering and Mining Journal, April 14, 1906; April 21, 1906; May 30, 1908; April 24, 1909; October 15, 1910; July 8, 1911; September 30, 1911; from February 17, 1912, to May 31, 1913; August 19, 1916; June 23, 1917; January 22, 1916; January 5, 1918; February 7, 1914; March 28, 1914; April 4, 1914; August 22, 1914.

Black Sands of California. Engineering and Mining Journal, June 1, 1907; August 10, 1907; February 8, 1908.

Nomenclature of Placers. Engineering and Mining Journal, August 17, 1907.

Brandy City Mine. Engineering and Mining Journal, June 4, 1910.

Trinity River Gravel. Engineering and Mining Journal, September 9, 1911; May 17, 1913.

Drift Mining in California. Engineering and Mining Journal, October 23, 1915; November 17, 1917.

Gravel Mines of the Sierras. Mineral Resources West of Rocky Mountains for years 1871, 1872, 1873 and 1874.

Tertiary Gravels of the Sierras. Professional Paper No. 73, U. S. G. S.

Placerville and Georgetown Region. Mineral Resources West of the Rocky Mountains, 1872.

Placer Mines of Nevada and Butte Counties. Mineral Resources West of the Rocky Mountains, 1872.

Mines of Northern California. Mineral Resources West of the Rocky Mountains for the year 1875.

Mines of Plumas and Nevada Counties. Mineral Resources West of the Rocky Mountains for the year 1876.

Auriferous Gravels of the Sierras. Memoirs of the Museum of Comparative Zoology of Harvard; Contributions to American Geology, Vol. 1, by J. D. Whitney.

Hydraulic Mining Ditches of the Sierras. Mineral Resources West of the Rocky Mountains for the year 1868.

Study of the Forest Hill Divide. Tenth Annual Report, California State Mining Bureau.

Auriferous Black Sands of California. Bull. 45, California State Mining Bureau.

A Practical Treatise on Hydraulic Mining in California, by A. J. Bowie.

Reports of the State Mineralogist, California State Mining Bureau, from Vol. 1 to 17, inclusive. Especially the Fourteenth and Fifteenth Annual Reports.

Mining in California. California State Mining Bureau, 1922–1923.

Geology of the Lassen Peak District. Eighth Annual Report of the U. S. Geological Survey.

Drift Mining in California. Eighth Annual Report, California State Mining Bureau.

The Auriferous Gravels of California. Ninth Annual Report, California State Mining Bureau.

Auriferous Gravels of the Sierras. Folios Nos. 3, 5, 17, 18, 29, 31, 37, 39, 41, 43, 51 and 66 of the Geological Atlas of the United States, of the U. S. Geological Survey.

Ancient Channel System of Calaveras County. Twelfth Annual Report, California State Mining Bureau.

Gold Dredging in California. Bulletins Nos. 36 and 57, California State Mining Bureau.

Gold Dredging. Bulletin No. 127, United States Bureau of Mines.

Platinum and Allied Metals in California. Bull. 85, California State Mining Bureau.

INDEX.

O

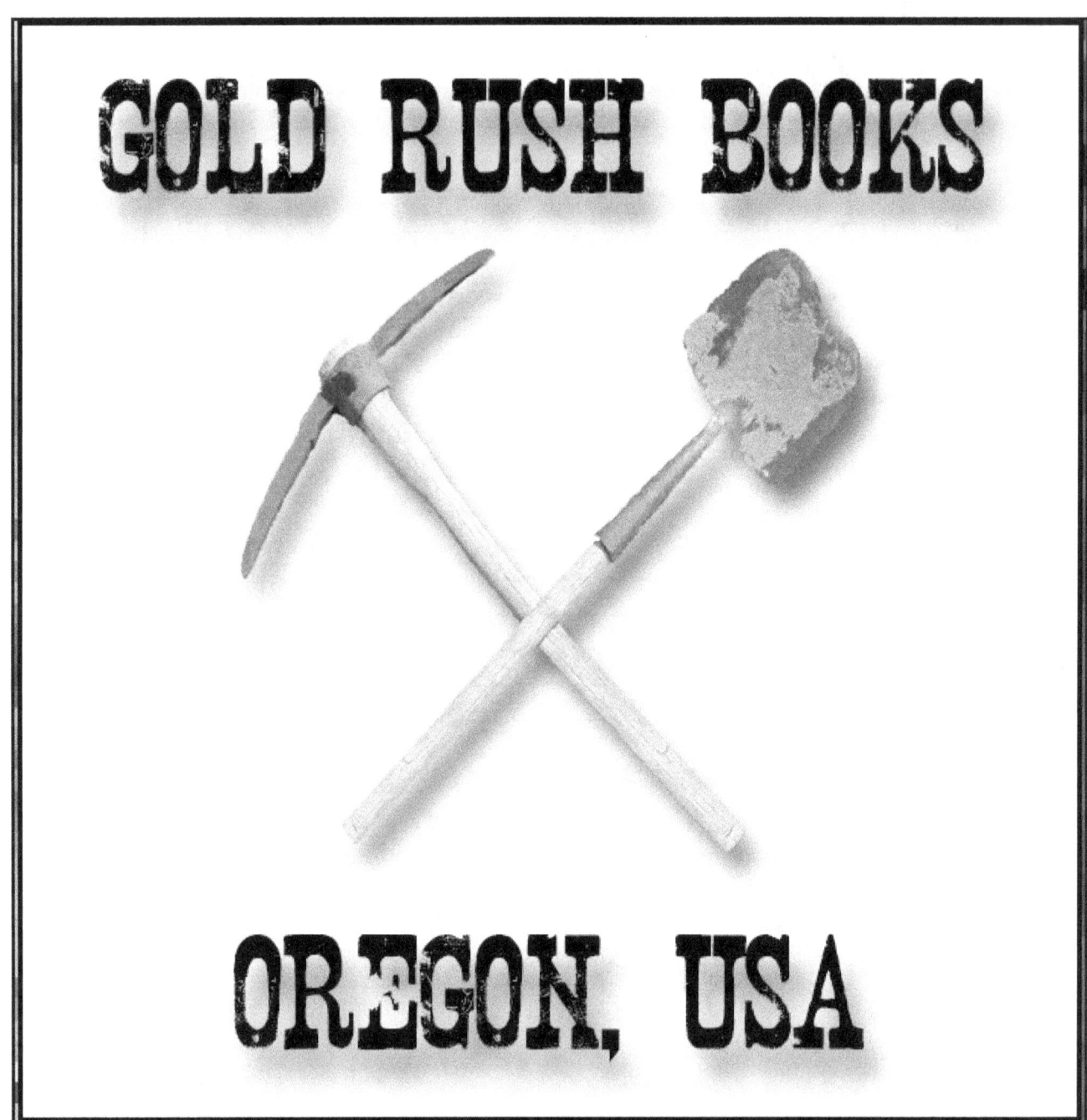

GOLD RUSH BOOKS

OREGON, USA

www.GoldMiningBooks.com

Books On Mining

Visit: www.goldminingbooks.com to order your copies
or ask your favorite book seller to offer them.

Mining Books by Kerby Jackson

Gold Dust: Stories From Oregon's Mining Years - Oregon mining historian and prospector, Kerby Jackson, brings you a treasure trove of seventeen stories on Southern Oregon's rich history of gold prospecting, the prospectors and their discoveries, and the breathtaking areas they settled in and made homes. 5" X 8", 98 ppgs. Retail Price: $11.99

The Golden Trail: More Stories From Oregon's Mining Years - In his follow-up to "Gold Dust: Stories of Oregon's Mining Years", this time around, Jackson brings us twelve tales from Oregon's Gold Rush, including the story about the first gold strike on Canyon Creek in Grant County, about the old timers who found gold by the pail full at the Victor Mine near Galice, how Iradel Bray discovered a rich ledge of gold on the Coquille River during the height of the Rogue River War, a tale of two elderly miners on the hunt for a lost mine in the Cascade Mountains, details about the discovery of the famous Armstrong Nugget and others. 5" X 8", 70 ppgs. Retail Price: $10.99

Oregon Mining Books

Geology and Mineral Resources of Josephine County, Oregon - Unavailable since the 1970's, this important publication was originally compiled by the Oregon Department of Geology and Mineral Industries and includes important details on the economic geology and mineral resources of this important mining area in South Western Oregon. Included are notes on the history, geology and development of important mines, as well as insights into the mining of gold, copper, nickel, limestone, chromium and other minerals found in large quantities in Josephine County, Oregon. 8.5" X 11", 54 ppgs. Retail Price: $9.99

Mines and Prospects of the Mount Reuben Mining District - Unavailable since 1947, this important publication was originally compiled by geologist Elton Youngberg of the Oregon Department of Geology and Mineral Industries and includes detailed descriptions, histories and the geology of the Mount Reuben Mining District in Josephine County, Oregon. Included are notes on the history, geology, development and assay statistics, as well as underground maps of all the major mines and prospects in the vicinity of this much neglected mining district. 8.5" X 11", 48 ppgs. Retail Price: $9.99

The Granite Mining District - Notes on the history, geology and development of important mines in the well known Granite Mining District which is located in Grant County, Oregon. Some of the mines discussed include the Ajax, Blue Ribbon, Buffalo, Continental, Cougar-Independence, Magnolia, New York, Standard and the Tillicum. Also included are many rare maps pertaining to the mines in the area. 8.5" X 11", 48 ppgs. Retail Price: $9.99

Ore Deposits of the Takilma and Waldo Mining Districts of Josephine County, Oregon - The Waldo and Takilma mining districts are most notable for the fact that the earliest large scale mining of placer gold and copper in Oregon took place in these two areas. Included are details about some of the earliest large gold mines in the state such as the Llano de Oro, High Gravel, Cameron, Platerica, Deep Gravel and others, as well as copper mines such as the famous Queen of Bronze mine, the Waldo, Lily and Cowboy mines. This volume also includes six maps and 20 original illustrations. 8.5" X 11", 74 ppgs. Retail Price: $9.99

Metal Mines of Douglas, Coos and Curry Counties, Oregon - Oregon mining historian Kerby Jackson introduces us to a classic work on Oregon's mining history in this important re-issue of Bulletin 14C Volume 1, otherwise known as the Douglas, Coos & Curry Counties, Oregon Metal Mines Handbook. Unavailable since 1940, this important publication was originally compiled by the Oregon Department of Geology and Mineral Industries includes detailed descriptions, histories and the geology of over 250 metallic mineral mines and prospects in this rugged area of South West Oregon. 8.5" X 11", 158 ppgs. Retail Price: $19.99

Metal Mines of Jackson County, Oregon - Unavailable since 1943, this important publication was originally compiled by the Oregon Department of Geology and Mineral Industries includes detailed descriptions, histories and the geology of over 450 metallic mineral mines and prospects in Jackson County, Oregon. Included are such famous gold mining areas as Gold Hill, Jacksonville, Sterling and the Upper Applegate. **8.5" X 11", 220 ppgs. Retail Price: $24.99**

Metal Mines of Josephine County, Oregon - Oregon mining historian Kerby Jackson introduces us to a classic work on Oregon's mining history in this important re-issue of Bulletin 14C, otherwise known as the Josephine County, Oregon Metal Mines Handbook. Unavailable since 1952, this important publication was originally compiled by the Oregon Department of Geology and Mineral Industries includes detailed descriptions, histories and the geology of over 500 metallic mineral mines and prospects in Josephine County, Oregon. **8.5" X 11", 250 ppgs. Retail Price: $24.99**

Metal Mines of North East Oregon - Oregon mining historian Kerby Jackson introduces us to a classic work on Oregon's mining history in this important re-issue of Bulletin 14A and 14B, otherwise known as the North East Oregon Metal Mines Handbook. Unavailable since 1941, this important publication was originally compiled by the Oregon Department of Geology and Mineral Industries and includes detailed descriptions, histories and the geology of over 750 metallic mineral mines and prospects in North Eastern Oregon. **8.5" X 11", 310 ppgs. Retail Price: $29.99**

Metal Mines of North West Oregon - Oregon mining historian Kerby Jackson introduces us to a classic work on Oregon's mining history in this important re-issue of Bulletin 14D, otherwise known as the North West Oregon Metal Mines Handbook. Unavailable since 1951, this important publication was originally compiled by the Oregon Department of Geology and Mineral Industries and includes detailed descriptions, histories and the geology of over 250 metallic mineral mines and prospects in North Western Oregon. **8.5" X 11", 182 ppgs. Retail Price: $19.99**

Mines and Prospects of Oregon - Mining historian Kerby Jackson introduces us to a classic mining work by the Oregon Bureau of Mines in this important re-issue of The Handbook of Mines and Prospects of Oregon. Unavailable since 1916, this publication includes important insights into hundreds of gold, silver, copper, coal, limestone and other mines that operated in the State of Oregon around the turn of the 19th Century. Included are not only geological details on early mines throughout Oregon, but also insights into their history, production, locations and in some cases, also included are rare maps of their underground workings. **8.5" X 11", 314 ppgs. Retail Price: $24.99**

Lode Gold of the Klamath Mountains of Northern California and South West Oregon
(See California Mining Books)

Mineral Resources of South West Oregon - Unavailable since 1914, this publication includes important insights into dozens of mines that once operated in South West Oregon, including the famous gold fields of Josephine and Jackson Counties, as well as the Coal Mines of Coos County. Included are not only geological details on early mines throughout South West Oregon, but also insights into their history, production and locations. **8.5" X 11", 154 ppgs. Retail Price: $11.99**

Chromite Mining in The Klamath Mountains of California and Oregon
(See California Mining Books)

Southern Oregon Mineral Wealth - Unavailable since 1904, this rare publication provides a unique snapshot into the mines that were operating in the area at the time. Included are not only geological details on early mines throughout South West Oregon, but also insights into their history, production and locations. Some of the mining areas include Grave Creek, Greenback, Wolf Creek, Jump Off Joe Creek, Granite Hill, Galice, Mount Reuben, Gold Hill, Galls Creek, Kane Creek, Sardine Creek, Birdseye Creek, Evans Creek, Foots Creek, Jacksonville, Ashland, the Applegate River, Waldo, Kerby and the Illinois River, Althouse and Sucker Creek, as well as insights into local copper mining and other topics. **8.5" X 11", 64 ppgs. Retail Price: $8.99**

Geology and Ore Deposits of the Takilma and Waldo Mining Districts - Unavailable since the 1933, this publication was originally compiled by the United States Geological Survey and includes details on gold and copper mining in the Takilma and Waldo Districts of Josephine County, Oregon. The Waldo and Takilma mining districts are most notable for the fact that the earliest large scale mining of placer gold and copper in Oregon took place in these two areas. Included in this report are details about some of the earliest large gold mines in the state such as the Llano de Oro, High Gravel, Cameron, Platerica, Deep Gravel and others, as well as copper mines such as the famous Queen of Bronze mine, the Waldo, Lily and Cowboy mines. In addition to geological examinations, insights are also provided into the production, day to day operations and early histories of these mines, as well as calculations of known mineral reserves in the area. This volume also includes six maps and 20 original illustrations. **8.5" X 11", 74 ppgs. Retail Price: $9.99**

Gold Mines of Oregon - Oregon mining historian Kerby Jackson introduces us to a classic work on Oregon's mining history in this important re-issue of Bulletin 61, otherwise known as "Gold and Silver In Oregon". Unavailable since 1968, this important publication was originally compiled by geologists Howard C. Brooks and Len Ramp of the Oregon Department of Geology and Mineral Industries and includes detailed descriptions, histories and the geology of over 450 gold mines Oregon. Included are notes on the history, geology and gold production statistics of all the major mining areas in Oregon including the Klamath Mountains, the Blue Mountains and the North Cascades. While gold is where you find it, as every miner knows, the path to success is to prospect for gold where it was previously found. 8.5" X 11", 344 ppgs. **Retail Price: $24.99**

Mines and Mineral Resources of Curry County Oregon - Originally published in 1916, this important publication on Oregon Mining has not been available for nearly a century. Included are rare insights into the history, production and locations of dozens of gold mines in Curry County, Oregon, as well as detailed information on important Oregon mining districts in that area such as those at Agness, Bald Face Creek, Mule Creek, Boulder Creek, China Diggings, Collier Creek, Elk River, Gold Beach, Rock Creek, Sixes River and elsewhere. Particular attention is especially paid to the famous beach gold deposits of this portion of the Oregon Coast. 8.5" X 11", 140 ppgs. **Retail Price: $11.99**

Chromite Mining in South West Oregon - Originally published in 1961, this important publication on Oregon Mining has not been available for nearly a century. Included are rare insights into the history, production and locations of nearly 300 chromite mines in South Western Oregon. 8.5" X 11", 184 ppgs. **Retail Price: $14.99**

Mineral Resources of Douglas County Oregon - Originally published in 1972, this important publication on Oregon Mining has not been available for nearly forty years. Included are rare insights into the geology, history, production and locations of numerous gold mines and other mining properties in Douglas County, Oregon. 8.5" X 11", 124 ppgs. **Retail Price: $11.99**

Mineral Resources of Coos County Oregon - Originally published in 1972, this important publication on Oregon Mining has not been available for nearly forty years. Included are rare insights into the geology, history, production and locations of numerous gold mines and other mining properties in Coos County, Oregon. 8.5" X 11", 100 ppgs. **Retail Price: $11.99**

Mineral Resources of Lane County Oregon - Originally published in 1938, this important publication on Oregon Mining has not been available for nearly seventy five years. Included are extremely rare insights into the geology and mines of Lane County, Oregon, in particular in the Bohemia, Blue River, Oakridge, Black Butte and Winberry Mining Districts. 8.5" X 11", 82 ppgs. **Retail Price: $9.99**

Mineral Resources of the Upper Chetco River of Oregon: Including the Kalmiopsis Wilderness - Originally published in 1975, this important publication on Oregon Mining has not been available for nearly forty years. Withdrawn under the 1872 Mining Act since 1984, real insight into the minerals resources and mines of the Upper Chetco River has long been unavailable due to the remoteness of the area. Despite this, the decades of battle between property owners and environmental extremists over the last private mining inholding in the area has continued to pique the interest of those interested in mining and other forms of natural resource use. Gold mining began in the area in the 1850's and has a rich history in this geographic area, even if the facts surrounding it are little known. Included are twenty two rare photographs, as well as insights into the Becca and Morning Mine, the Emmly Mine (also known as Emily Camp), the Frazier Mine, the Golden Dream or Higgins Mine, Hustis Mine, Peck Mine and others. 8.5" X 11", 64 ppgs. **Retail Price: $8.99**

Gold Dredging in Oregon - Originally published in 1939, this important publication on Oregon Mining has not been available for nearly seventy five years. Included are extremely rare insights into the history and day to day operations of the dragline and bucketline gold dredges that once worked the placer gold fields of South West and North East Oregon in decades gone by. Also included are details into the areas that were worked by gold dredges in Josephine, Jackson, Baker and Grant counties, as well as the economic factors that impacted this mining method. This volume also offers a unique look into the values of river bottom land in relation to both farming and mining, in how farm lands were mined, re-soiled and reclamated after the dredges worked them. Featured are hard to find maps of the gold dredge fields, as well as rare photographs from a bygone era. 8.5" X 11", 86 ppgs. **Retail Price: $8.99**

Quick Silver Mining in Oregon - Originally published in 1963, this important publication on Oregon Mining has not been available for over fifty years. This publication includes details into the history and production of Elemental Mercury or Quicksilver in the State of Oregon. 8.5" X 11", 238 ppgs. **Retail Price: $15.99**

Mines of the Greenhorn Mining District of Grant County Oregon - Originally published in 1948, this important publication on Oregon Mining has not been available for over sixty five years. In this publication are rare insights into the mines of the famous Greenhorn Mining District of Grant County, Oregon, especially the famous Morning Mine. Also included are details on the Tempest, Tiger, Bi-Metallic, Windsor, Psyche, Big Johnny, Snow Creek, Banzette and Paramount Mines, as well as prospects in the vicinities in the famous mining areas of Mormon Basin, Vinegar Basin and Desolation Creek. Included are hard to find mine maps and dozens of rare photographs from the bygone era of Grant County's rich mining history. 8.5" X 11", 72 ppgs. **Retail Price: $9.99**

Geology of the Wallowa Mountains of Oregon: Part I (Volume 1) - Originally published in 1938, this important publication on Oregon Mining has not been available for nearly seventy five years. Included are details on the geology of this unique portion of North Eastern Oregon. This is the first part of a two book series on the area. Accompanying the text are rare photographs and historic maps.**8.5" X 11", 92 ppgs. Retail Price: $9.99**

Geology of the Wallowa Mountains of Oregon: Part II (Volume 2) - Originally published in 1938, this important publication on Oregon Mining has not been available for nearly seventy five years. Included are details on the geology of this unique portion of North Eastern Oregon. This is the first part of a two book series on the area. Accompanying the text are rare photographs and historic maps.**8.5" X 11", 94 ppgs. Retail Price: $9.99**

Field Identification of Minerals For Oregon Prospectors - Originally published in 1940, this important publication on Oregon Mining has not been available for nearly seventy five years. Included in this volume is an easy system for testing and identifying a wide range of minerals that might be found by prospectors, geologists and rockhounds in the State of Oregon, as well as in other locales. Topics include how to put together your own field testing kit and how to conduct rudimentary tests in the field. This volume is written in a clear and concise way to make it useful even for beginners. **8.5" X 11", 158 ppgs. Retail Price: $14.99**

The Bohemia Mining District of Oregon - Originally published in 1900, this important publication on Oregon Mining has not been available for over a century. Included in this volume are important insights into the famous Bohemia Mining District of Oregon, including the histories and locations of important gold mines in the area such as the Ophir Mine, Clarence, Acturas, Peek-a-boo, White Swan, Combination Mine, the Musick Mine, The California, White Ghost, The Mystery, Wall Street, Vesuvius, Story, Lizzie Bullock, Delta, Elsie Dora, Golden Slipper, Broadway, Champion Mine, Knott, Noonday, Helena, White Wings, Riverside and others. Also included are notes on the nearby Blue River Mining District. **8.5" X 11", 58 ppgs. Retail Price: $9.99**

The Gold Fields of Eastern Oregon - Unavailable since 1900, this publication was originally compiled by the Baker City Chamber of Commerce Offering important insights into the gold mining history of Eastern Oregon, "The Gold Fields of Eastern Oregon" sheds a rare light on many of the gold mines that were operating at the turn of the 19th Century in Baker County and Grant County in North Eastern Oregon. Some of the areas featured include the Cable Cove District, Baisely-Elhorn, Granite, Red Boy, Bonanza, Susanville, Sparta, Virtue, Vaughn, Sumpter, Burnt River, Rye Valley and other mining districts. Included is basic information on not only many gold mines that are well known to those interested in Eastern Oregon mining history, but also many mines and prospects which have been mostly lost to the passage of time. Accompanying are numerous rare photos **8.5" X 11", 78 ppgs. Retail Price: $10.99**

Gold Mining in Eastern Oregon - Originally published in 1938, this important publication on Oregon Mining has not been available for over a century. Included in this volume are important insights into the famous mining districts of Eastern Oregon during the late 1930's. Particular attention is given to those gold mines with milling and concentrating facilities in the Greenhorn, Red Boy, Alamo, Bonanza, Granite, Cable Cove, Cracker Creek, Virtue, Keating, Medical Springs, Sanger, Sparta, Chicken Creek, Mormon Basin, Connor Creek, Cornucopia and the Bull Run Mining Districts. Some of the mines featured include the Ben Harrison, North Pole-Columbia, Highland Maxwell, Baisley-Elkhorn, White Swan, Balm Creek, Twin Baby, Gem of Sparta, New Deal, Gleason, Gifford-Johnson, Cornucopia, Record, Bull Run, Orion and others. Of particular interest are the mill flow sheets and descriptions of milling operations of these mines. **8.5" X 11", 68 ppgs. Retail Price: $8.99**

The Gold Belt of the Blue Mountains of Oregon - Originally published in 1901, this important publication on Oregon Mining has not been available for over a century. Included in this volume are rare insights into the gold deposits of the Blue Mountains of North East Oregon, including the history of their early discovery and early production. Extensive details are offered on this important mining area's mineralogy and economic geology, as well as insights into nearby gold placers, silver deposits and copper deposits. Featured are the Elkhorn and Rock Creek mining districts, the Pocahontas district, Auburn and Minersville districts, Sumpter and Cracker Creek, Cable Cove, the Camp Carson district, Granite, Alamo, Greenhorn, Robinsonville, the Upper Burnt River Valley and Bonanza districts, Susanville, Quartzburg, Canyon Creek, Virtue, the Copper Butte district, the North Powder River, Sparta, Eagle Creek, Cornucopia, Pine Creek, Lower Powder River, the Upper Snake River Canyon, Rye Valley, Lower Burnt River Valley, Mormon Basin, the Malheur and Clarks Creek districts, Sutton Creek and others. Of particular interest are important details on numerous gold mines and prospects in these mining districts, including their locations, histories, geology and other important information, as well as information on silver, copper and fire opal deposits. **8.5" X 11", 250 ppgs. Retail Price: $24.99**

Mining in the Cascades Range of Oregon - Originally published in 1938, this important publication on Oregon Mining has not been available for over seventy five years. Included in this volume are rare insights into the gold mines and other types of metal mines in the Cascades Mountain Range of Oregon. Some of the important mining areas covered include the famous Bohemia Mining District, the North Santiam Mining District, Quartzville Mining District, Blue River Mining District, Fall Creek Mining District, Oakridge District, Zinc District, Buzzard-Al Sarena District, Grand Cove, Climax District and Barron Mining District. Of particular interest are important details on over 100 mines and prospects in these mining districts, including their locations, histories, geology and other important information. **8.5" X 11", 170 ppgs. Retail Price: $14.99**

Beach Gold Placers of the Oregon Coast - Originally published in 1934, this important publication on Oregon Mining has not been available for over 80 years. Included in this volume are rare insights into the beach gold deposits of the State of Oregon, including their locations, occurance, composition and geology. Of particular interest is information on placer platinum in Oregon's rich beach deposits. Also included are the locations and other information on some famous Oregon beach mines, including the Pioneer, Eagle, Chickamin, Iowa and beach placer mines north of the mouth of the Rogue River. **8.5" X 11", 60 ppgs. Retail Price: $8.99**

Mineralogical Composition of the Sands of the Oregon Coast: From Coos Bay to the Columbia - Published in 1945, he text features hard to find information on the composition of the gold bearing black sands of the South West Oregon Coast, offering a unique insight to prospectors in search of Oregon's legendary beach gold. 104 ppgs, $9.99

Manganese Mining in Oregon - First released in 1942 and now out of print, this special reprint edition of "Manganese in Oregon" was originally published by the Oregon Department of Geology and Mineral Industries. The text features hard to find information on the mining of Manganese in Oregon, including details and maps of Oregon manganese mines and prospects. 108 ppgs, 9.99

Medford Oregon As A Mining Center - Written in 1912, this hard to find publication includes valuable insights into the mining history of South West Oregon. This small book contains interesting information on the gold, copper and mining industry in Southern Oregon as it existed just prior to World War One, shedding light on some of the important mines in the area. Included are rare photographs and vintage advertising of the day. 80 ppgs, 9.99

Mineral Resources of Curry County Oregon - First released in 1977 and now out of print, this special reprint edition of "Geology, Mineral Resources and Rock Materials of Curry County, Oregon" was originally published in cooperation of Curry County, Oregon and the Oregon Department of Geology and Mineral Industries. The text features hard to find information on not only the mining of gold and other metals in Curry County, but also aggregate mining in the area. 102 ppgs, 11.99

Origin of the Gold Bearing Black Sands of the Coast of South West Oregon - First released in 1943 and now out of print, this special reprint edition of "The Origin of the Black Sands of the South West Oregon Coast" was originally published by the Oregon Department of Geology and Mineral Industries. The text features hard to find information on the origin of the gold bearing black sands of the South West Oregon Coast, offering a unique insight to prospectors in search of Oregon's legendary beach gold. 52 ppgs, 8.99

South West Oregon Mining - Leading mining historian Kerby Jackson introduces us to six classic small mining publications on the Gold Mining Industry in Southern Oregon. This small book consists of a compilation of USGS J.S. Diller's "Mines of the Riddles Quadrangle", "The Rogue River Valley Coal Fields" and "Mineral Resources of the Grants Pass Quadrangle", the Grants Pass Commercial Club's rare publication "Mining in Josephine County, Oregon" and the USGS publication "The Distribution of Placer Gold in the Sixes River, South West Oregon". Also included is F.W. Libbey's legendary article on the Southern Oregon Mining Industry, "Lest We Forget", which appeared in the publication of the Oregon State Department of Geology and Mineral Industries in the early 1960's. This compilation offers a unique perspective on mining in South West Oregon and includes considerable information on mines in Josephine, Jackson and Coos Counties. 142 ppgs, 14.99

Geology and Mineral Resources of the Gasquet Quadrangle of California-Oregon - First published in 1953, it has been unavailable for over a century and sheds important light on the geological features and mineral resources of this portion of Northern California and Southern Oregon. 80 ppgs, 9.99

Idaho Mining Books

Gold in Idaho - Unavailable since the 1940's, this publication was originally compiled by the Idaho Bureau of Mines and includes details on gold mining in Idaho. Included is not only raw data on gold production in Idaho, but also valuable insight into where gold may be found in Idaho, as well as practical information on the gold bearing rocks and other geological features that will assist those looking for placer and lode gold in the State of Idaho. This volume also includes thirteen gold maps that greatly enhance the practical usability of the information contained in this small book detailing where to find gold in Idaho. **8.5″ X 11″, 72 ppgs. Retail Price: $9.99**

Geology of the Couer D'Alene Mining District of Idaho - Unavailable since 1961, this publication was originally compiled by the Idaho Bureau of Mines and Geology and includes details on the mining of gold, silver and other minerals in the famous Coeur D'Alene Mining District in Northern Idaho. Included are details on the early history of the Coeur D'Alene Mining District, local tectonic settings, ore deposit features, information on the mineral belts of the Osburn Fault, as well as detailed information on the famous Bunker Hill Mine, the Dayrock Mine, Galena Mine, Lucky Friday Mine and the infamous Sunshine Mine. This volume also includes sixteen hard to find maps. **8.5″ X 11″, 70 ppgs. Retail Price: $9.99**

The Gold Camps and Silver Cities of Idaho - Originally published in 1963, this important publication on Idaho Mining has not been available for nearly fifty years. Included are rare insights into the history of Idaho's Gold Rush, as well as the mad craze for silver in the Idaho Panhandle. Documented in fine detail are the early mining excitements at Boise Basin, at South Boise, in the Owyhees, at Deadwood, Long Valley, Stanley Basin and Robinson Bar, at Atlanta, on the famous Boise River, Volcano, Little Smokey, Banner, Boise Ridge, Hailey, Leesburg, Lemhi, Pearl, at South Mountain, Shoup and Ulysses, Yellow Jacket and Loon Creek. The story follows with the appearance of Chinese miners at the new mining camps on the Snake River, Black Pine, Yankee Fork, Bay Horse, Clayton, Heath, Seven Devils, Gibbonsville, Vienna and Sawtooth City. Also included are special sections on the Idaho Lead and Silver mines of the late 1800's, as well as the mining discoveries of the early 1900's that paved the way for Idaho's modern mining and mineral industry. Lavishly illustrated with rare historic photos, this volume provides a one of a kind documentary into Idaho's mining history that is sure to be enjoyed by not only modern miners and prospectors who still scour the hills in search of nature's treasures, but also those enjoy history and tromping through overgrown ghost towns and long abandoned mining camps. **8.5″ X 11″, 186 ppgs. Retail Price: $14.99**

Ore Deposits and Mining in North Western Custer County Idaho - Unavailable since 1913, this important publication was originally published by the Us Department of the Interior and has been unavailable for a century. Included are fine details on the geology, geography, gold placers and gold and silver bearing quartz veins of the mining region of North West Custer County, Idaho. Of particular interest is a rare look at the mines and prospects of the region, including those such as the Ramshorn Mine, SkyLark, Riverview, Excelsior, Beardsley, Pacific, Hoosier, Silver Brick, Forest Rose and dozens of others in the Bay Horse Mining District. Also covered are the mines of the Yankee Fork District such as the Lucky Boy, Badger, Black, Enterprise, Charles Dickens, Morrison, Golden Sunbeam, Montana, Golden Gate and others, as well as those in the Loon Mining District. **8.5″ X 11″, 126 ppgs. Retail Price: $12.99**

Gold Rush To Idaho - Unavailable since 1963, this important publication was originally published by the Idaho Bureau of Mines and has been unavailable for 50 years. "Gold Rush To Idaho" revisits the earliest years of the discovery of gold in Idaho Territory and introduces us to the conditions that the pioneer gold seekers met when they blazed a trail through the wilderness of Idaho's mountains and discovered the precious yellow metal at Oro Fino and Pierce. Subsequent rushes followed at places like Elk City, Newsome, Clearwater Station, Florence, Warrens and elsewhere. Of particular interest is a rare look at the hardships that the first miners in Idaho met with during their day to day existences and their attempts to bring law and order to their mining camps. **8.5″ X 11″, 88 ppgs. Retail Price: $9.99**

The Geology and Mines of Northern Idaho and North Western Montana - Unavailable since 1909, this important publication was originally published by the Us Department of the Interior and has been unavailable for a century. Included are fine details on the geology and geography of the mining regions of Northern Idaho and North Western Montana. Of particular interest is a rare look at the mines and prospects of the region, including those in the Pine Creek Mining District, Lake Pend Oreille district, Troy Mining District, Sylvanite District, Cabinet Mining District, Prospect Mining District and the Missoula Valley. Some of the mines featured include the Iron Mountain, Silver Butte, Snowshoe, Grouse Mountain Mine and others. **8.5″ X 11″, 142 ppgs. Retail Price: $12.99**

Mining in the Alturas Quadrangle of Blaine County Idaho - Unavailable since 1922, this important publication was originally published by the Idaho Bureau of Mines and has been unavailable for ninety years. Topics include the geology, rock formations and the formation of ore deposits in this important mining area of Idaho. Of particular focus is information on the local geology, quartz veins and ore deposits of this portion of Idaho. Included are hard to find details, including the descriptions and locations of numerous gold and silver mines in the area including the Silver King, Pilgrim, Columbia, Lone Jack, Sunbeam, Pride of the West, Lucky Boy, Scotia, Atlanta, Beaver-Bidwell and others mines and prospects. **8.5″ X 11″, 56 ppgs. Retail Price: $8.99**

Mining in Lemhi County Idaho - Originally published in 1913, this important book on Idaho Mining has not been available to miners for over a century. Included are rare insights into hundreds of gold, silver, copper and other mines in this famous Idaho mining area. Details include the locations, geology, history, production and other facts of the mines of this region, not only gold and silver hardrock mines, but also gold placer mines, lead-silver deposits, copper mines, cobalt-nickel deposits, tungsten and tin mines . It is lavishly illustrated with hard to find photos of the period and rare mining maps. Some of the vicinities featured include the Nicholia Mining District, Spring Mountain District, Texas District, Blue Wing District, Junction District, McDevitt District, Pratt Creek, Eldorado District, Kirtley Creek, Carmen Creek, Gibbonsville, Indian Creek, Mineral Hill District, Mackinaw, Eureka District, Blackbird District, YellowJacket District, Gravel Range District, Junction District, Parker Mountain and other mining districts. **8.5" X 11", 226 ppgs. Retail Price: $19.99**

Mining in Shoshone County Idaho - First published in 1923, it has been unavailable for over a century and sheds important light on the mining history of Shoshone County, Idaho. Some of the topics include the history of mining in Shoshone County, a look at the local geology and ore characteristics of lead-silver deposits, zinc deposits, copper, antimony, gold and other minerals. Also included are insights into the history, production, characteristics and locations of numerous mines in the area. 198 ppgs, 15.99

Utah Mining Books

Fluorite in Utah – Unavailable since 1954, this publication was originally compiled by the USGS, State of Utah and U.S. Atomic Energy Commission and details the mining of fluorspar, also known as fluorite in the State of Utah. Included are details on the geology and history of fluorspar (fluorite) mining in Utah, including details on where this unique gem mineral may be found in the State of Utah. **8.5" X 11", 60 ppgs. Retail Price: $8.99**

The Gold Hill Mining District of Utah - First published in 1935, it has been unavailable since those days and sheds important light on the mines, history and geology of Utah's Gold Hill Mining District. Included are rare insights into this important mining area, including the locations, histories and details of numerous mines. This volume is well illustrated with geological diagrams, as well as hard to find maps of some of the most important mines in this district. 202 ppgs., 19.99

The Mines, Miners and Minerals of Utah - First published in 1896, it has been unavailable since those days and sheds important light on the early mines and miners of Pioneer Utah, as well as the minerals which they won from the earth by laborious hard physical labor and sheer determination. Included are rare insights into the early mining history of Utah, as well details on hundreds of gold, silver and copper mines. 376 ppgs., 24.99

California Mining Books

The Tertiary Gravels of the Sierra Nevada of California – Mining historian Kerby Jackson introduces us to a classic mining work by Waldemar Lindgren in this important re-issue of The Tertiary Gravels of the Sierra Nevada of California. Unavailable since 1911, this publication includes details on the gold bearing ancient river channels of the famous Sierra Nevada region of California. **8.5" X 11", 282 ppgs. Retail Price: $19.99**

The Mother Lode Mining Region of California – Unavailable since 1900, this publication includes details on the gold mines of California's famous Mother Lode gold mining area. Included are details on the geology, history and important gold mines of the region, as well as insights into historic mining methods, mine timbering, mining machinery, mining bell signals and other details on how these mines operated. Also included are insights into the gold mines of the California Mother Lode that were in operation during the first sixty years of California's mining history. **8.5" X 11", 176 ppgs. Retail Price: $14.99**

Lode Gold of the Klamath Mountains of Northern California and South West Oregon – Unavailable since 1971, this publication was originally compiled by Preston E. Hotz and includes details on the lode mining districts of Oregon and California's Klamath Mountains. Included are details on the geology, history and important lode mines of the French Gulch, Deadwood, Whiskeytown, Shasta, Redding, Muletown, South Fork, Old Diggings, Dog Creek (Delta), Bully Choop (Indian Creek), Harrison Gulch, Hayfork, Minersville, Trinity Center, Canyon Creek, East Fork, New River, Denny, Liberty (Black Bear), Cecilville, Callahan, Yreka, Fort Jones and Happy Camp mining districts in California, as well as the Ashland, Rogue River, Applegate, Illinois River, Takilma, Greenback, Galice, Silver Peak, Myrtle Creek and Mule Creek districts of South Western Oregon. Also included are insights into the mineralization and other characteristics of this important mining region. **8.5" X 11", 100 ppgs. Retail Price: $10.99**

Mines and Mineral Resources of Shasta County, Siskiyou County, Trinity County: California – Unavailable since 1915, this publication was originally compiled by the California State Mining Bureau and includes details on the gold mines of this area of Northern California. Also included are insights into the mineralization and other characteristics of this important mining region, as well as the location of historic gold mines. **8.5" X 11", 204 ppgs. Retail Price: $19.99**

Geology of the Yreka Quadrangle, Siskiyou County, California - Unavailable since 1977, this publication was originally compiled by Preston E. Hotz and includes details on the geology of the Yreka Quadrangle of Siskiyou County, California. Also included are insights into the mineralization and other characteristics of this important mining region. 8.5" X 11", 78 ppgs. **Retail Price: $7.99**

Mines of San Diego and Imperial Counties, California - Originally published in 1914, this important publication on California Mining has not been available for a century. This publication includes important information on the early gold mines of San Diego and Imperial County, which were some of the first gold fields mined in California by early Spanish and Mexican miners before the 49ers came on the scene. Included are not only details on early mining methods in the area, production statistics and geological information, but also the location of the early gold mines that helped make California "The Golden State". Also included are details on the mining of other minerals such as silver, lead, zinc, manganese, tungsten, vanadium, asbestos, barite, borax, cement, clay, dolomite, fluospar, gem stones, graphite, marble, salines, petroleum, stronium, talc and others. 8.5" X 11", 116 ppgs. **Retail Price: $12.99**

Mines of Sierra County, California - Unavailable since 1920, this publication was originally compiled by the California State Mining Bureau and includes details on the gold mines of Sierra County, California. Also included are insights into the mineralization and other characteristics of this important mining region, as well as the location of historic gold mines. 8.5" X 11", 156 ppgs. **Retail Price: $19.99**

Mines of Plumas County, California - Unavailable since 1918, this publication was originally compiled by the California State Mining Bureau and includes details on the gold mines of Plumas County, California. Also included are insights into the mineralization and other characteristics of this important mining region, as well as the location of historic gold mines. 8.5" X 11", 200 ppgs. **Retail Price: $19.99**

Mines of El Dorado, Placer, Sacramento and Yuba Counties, California - Originally published in 1917, this important publication on California Mining has not been available for nearly a century. This publication includes important information on the early gold mines of El Dorado County, Placer County, Sacramento County and Yuba County, which were some of the first gold fields mined by the Forty-Niners during the California Gold Rush. Included are not only details on early mining methods in the area, production statistics and geological information, but also the location of the early gold mines that helped make California "The Golden State". Also included are insights into the early mining of chrome, copper and other minerals in this important mining area. 8.5" X 11", 204 ppgs. **Retail Price: $19.99**

Mines of Los Angeles, Orange and Riverside Counties, California - Originally published in 1917, this important publication on California Mining has not been available for nearly a century. This publication includes important information on the early gold mines of Los Angeles County, Orange County and Riverside County, which were some of the first gold fields mined in California by early Spanish and Mexican miners before the 49ers came on the scene. Included are not only details on early mining methods in the area, production statistics and geological information, but also the location of the early gold mines that helped make California "The Golden State". 8.5" X 11", 146 ppgs. **Retail Price: $12.99**

Mines of San Bernadino and Tulare Counties, California - Originally published in 1917, this important publication on California Mining has not been available for nearly a century. This publication includes important information on the early gold mines of San Bernadino and Tulare County, which were some of the first gold fields mined in California by early Spanish and Mexican miners before the 49ers came on the scene. Included are not only details on early mining methods in the area, production statistics and geological information, but also the location of the early gold mines that helped make California "The Golden State". Also included are details on the mining of other minerals such as copper, iron, lead, zinc, manganese, tungsten, vanadium, asbestos, barite, borax, cement, clay, dolomite, fluospar, gem stones, graphite, marble, salines, petroleum, stronium, talc and others. 8.5" X 11", 200 ppgs. **Retail Price: $19.99**

Chromite Mining in The Klamath Mountains of California and Oregon - Unavailable since 1919, this publication was originally compiled by J.S. Diller of the United States Department of Geological Survey and includes details on the chromite mines of this area of Northern California and Southern Oregon. Also included are insights into the mineralization and other characteristics of this important mining region, as well as the location of historic mines. Also included are insights into chromite mining in Eastern Oregon and Montana. 8.5" X 11", 98 ppgs. **Retail Price: $9.99**

Mines and Mining in Amador, Calaveras and Tuolumne Counties, California - Unavailable since 1915, this publication was originally compiled by William Tucker and includes details on the mines and mineral resources of this important California mining area. Included are details on the geology, history and important gold mines of the region, as well as insights into other local mineral resources such as asbestos, clay, copper, talc, limestone and others. Also included are insights into the mineralization and other characteristics of this important portion of California's Mother Lode mining region. 8.5" X 11", 198 ppgs. **Retail Price: $14.99**

The Cerro Gordo Mining District of Inyo County California - Unavailable since 1963, this publication was originally compiled by the United States Department of Interior. Included are insights into the mineralization and other characteristics of this important mining region of Southern California. Topics include the mining of gold and silver in this important mining district in Inyo County, California, including details on the history, production and locations of the Cerro Gordo Mine, the Morning Star Mine, Estelle Tunnel, Charles Lease Tunnel, Ignacio, Hart, Crosscut Tunnel, Sunset, Upper Newtown, Newtown, Ella, Perseverance, Newsboy, Belmont and other silver and gold mines in the Cerro Gordo Mining District. This volume also includes important insights into the fossil record, geologic formations, faults and other aspects of economic geology in this California mining district. **8.5" X 11", 104 ppgs. Retail Price: $10.99**

Mining in Butte, Lassen, Modoc, Sutter and Tehama Counties of California - Unavailable since 1917, this publication was originally compiled by the United States Department of Interior. Included are insights into the mineralization and other characteristics of this important mining region of California. Topics include the mining of asbestos, chromite, gold, diamonds and manganese in Butte County, the mining of gold and copper in the Hayden Hill and Diamond Mountain mining districts of Lassen County, the mining of coal, salt, copper and gold in the High Grade and Winters mining districts of Modoc County, gold mining in Sutter County and the mining of gold, chromite, manganese and copper in Tehama County. This volume also includes the production records and locations of numerous mines in this important mining region. **8.5" X 11", 114 ppgs. Retail Price: $11.99**

Mines of Trinity County California - Originally published in 1965, this important publication on California Mining has not been available for nearly fifty years. This publication includes important information on mines and mining in Trinity County, California, as well insights into the mineralization and geology of this important mining area in Northern California. Included are extensive details on hardrock and placer gold mines and prospects, including charts showing the locations of these historic mines.. **8.5" X 11", 144 ppgs. Retail Price: $12.99**

Mines of Kern County California - Originally published in 1962, this important publication on California Mining has not been available for nearly fifty years. This publication includes important information on mines and mining in Kern County, California, as well insights into the mineralization and geology of this important mining area in California. Included are extensive details on hardrock and placer gold mines and prospects, including charts showing the locations of these historic mines. **8.5" X 11", 398 ppgs. Retail Price: $24.99**

Mines of Calaveras County California - Originally published in 1962, this important publication on California Mining has not been available for nearly fifty years. This publication includes important information on mines and mining in Calaveras County, California, as well insights into the mineralization and geology of this important mining area in Northern California. Included are extensive details on hardrock and placer gold mines and prospects, including charts showing the locations of these historic mines. **8.5" X 11", 236 ppgs. Retail Price: $19.99**

Lode Gold Mining in Grass Valley California - Unavailable since 1940, this publication was originally compiled by the United States Department of Interior. Included are insights into the gold mineralization and other characteristics of this important mining region of Nevada County, California. This volume also includes important insights into the geologic formations, faults and other aspects of economic geology in this California mining district. Of particular interest are the fine details on many hardrock gold mines in the area, including their locations, histories, development and mineralization. Some of the mines featured include the Gold Hill Mine, Massachusetts Hill, Boundary, Peabody, Golden Center, North Star, Omaha, Lone Jack, Homeward Bound, Hartery, Wisconsin, Allison Ranch, Phoenix, Kate Hayes, W.Y.O.D., Empire, Rich Hill, Daisy Hill, Orleans, Sultana, Centennial, Conlin, Ben Franklin, Crown Point and many others. **8.5" X 11", 148 ppgs. Retail Price: $12.99**

Lode Mining in the Alleghany District of Sierra County California - Unavailable since 1913, this publication was originally compiled by the United States Department of Interior. Included are insights into the mineralization and other characteristics of this important mining region of Sierra County. Included are details on the history, production and locations of numerous hardrock gold mines in this famous California area, including the Tightner Mine, Minnie D., Osceola, Eldorado, Twenty One, Sherman, Kenton, Oriental, Rainbow, Plumbago, Irelan, Gold Canyon, North Fork, Federal, Kate Hardy and others. This volume also includes important insights into the fossil record, geologic formations, faults and other aspects of economic geology in this California mining district. **8.5" X 11", 48 ppgs. Retail Price: $7.99**

Six Months In The Gold Mines During The California Gold Rush - Unavailable since 1850, this important work is a first hand account of one "49'ers" personal experience during the great California Gold Rush, shedding important light on one of the most exciting periods in the history of not only California, but also the world. Compiled from journals written between 1847 and 1849 by E. Gould Buffum, a native of New York, "Six Months In The Gold Mines During The California Gold Rush" offers a rare look into the day to day lives of the people who came to California to work in her gold mines when the state was still a great frontier. **8.5" X 11", 290 ppgs. Retail Price: $19.99**

<u>Quartz Mines of the Grass Valley Mining District of California</u> - Unavailable since 1867, this important publication has not been available since those days. This rare publication offers a short dissertation on the early hardrock mines in this important mining district in the California Mother Lode region between the 1850's and 1860's. Also included are hard to find details on the mineralization and locations of these mines, as well as how they were operated in those day. 8.5" X 11", 44 ppgs. Retail Price: $8.99

<u>Gold Rush on the Feather River</u> - First published in 1924, this short publication by G.C. Mansfield sheds important light on the early history of gold mining on the Feather River. Included are rare insights into the first decade of gold mining and the early mining camps of the Feather River during the 1850's. 64 ppgs., 9.99

<u>The Bodie Mining District of California</u> - First published in 1986, it has been unavailable since those days and sheds important light on this famous mining area. Included are the history, characteristics and locations of numerous old mines around the ghost town of Bodie.
64 ppgs, 8.99

<u>Geology and Mineral Resources of the Gasquet Quadrangle of California-Oregon</u> - First published in 1953, it has been unavailable for over a century and sheds important light on the geological features and mineral resources of this portion of Northern California and Southern Oregon.
80 ppgs, 9.99

Alaska Mining Books

<u>Ore Deposits of the Willow Creek Mining District, Alaska</u> - Unavailable since 1954, this hard to find publication includes valuable insights into the Willow Creek Mining District near Hatcher Pass in Alaska. The publication includes insights into the history, geology and locations of the well known mines in the area, including the Gold Cord, Independence, Fern, Mabel, Lonesome, Snowbird, Schroff-O'Neil, High Grade, Marion Twin, Thorpe, Webfoot, Kelly-Willow, Lane, Holland and others. 8.5" X 11", 96 ppgs. Retail Price: $9.99

<u>The Juneau Gold Belt of Alaska</u> - Unavailable since 1906, this hard to find publication includes valuable insights into the gold mines around Juneau, Alaska. The publication includes important details into the history, geology and locations of the well known gold mines and prospects in the area, including those around Windham Bay, Holkham Bay, Port Snettisham, on Grindstone and Rhine Creeks, Gold Creek, Douglas Island, Salmon Creek, Lemon Creek, Nugget Creek, from the Mendenhall River to Berners Bay, McGinnis Creek, Montana Creek, Peterson Creek, Windfall Creek, the Eagle River, Yankee Basin, Yankee Curve, Kowee Creek and elsewhere. Not only are gold placer mines included, but also hardrock gold mines. 8.5" X 11", 224 ppgs. Retail Price: $19.99

<u>Mining in the Jumbo Basin of Alaska</u> - Unavailable since 1953, this hard to find publication includes valuable insights into the mines and geology of the Jumbo Basin. The publication includes important details into the history, geology and locations of the well known gold mines and prospects in the famous Jumbo Basin Mining Region of Alaska.
72 ppgs, 9.99

<u>The Rampart Placer Gold Region of Alaska</u> - Unavailable since 1906, this hard to find publication includes valuable insights into the placer gold mines of the Rampart Mining Region. The publication includes important details into the history, geology and locations of the well known gold mines and prospects in the famous Rampart Mining Region of Alaska.
78 ppgs, 10.99

Arizona Mining Books

<u>Mines and Mining in Northern Yuma County Arizona</u> - Originally published in 1911, this important publication on Arizona Mining has not been available for over a hundred years. Included are rare insights into the gold, silver, copper and quicksilver mines of Yuma County, Arizona together with hard to find maps and photographs. Some of the mines and mining districts featured include the Planet Copper Mine, Mineral Hill, the Clara Consolidated Mine, Viati Mine, Copper Basin prospect, Bowman Mine, Quartz King, Billy Mack, Carnation, the Wardwell and Osbourne, Valensuella Copper, the Mariquita, Colonial Mine, the French American, the New York-Plomosa, Guadalupe, Lead Camp, Mudersbach Copper Camp, Yellow Bird, the Arizona Northern (Salome Strike), Bonanza (Harqua Hala), Golden Eagle, Hercules, Socorro and others. 8.5" X 11", 144 ppgs. Retail Price: $11.99

<u>The Aravaipa and Stanley Mining Districts of Graham County Arizona</u> - Originally published in 1925, this important publication on Arizona Mining has not been available for nearly ninety years. Included are rare insights into the gold and silver mines of these two important mining districts, together with hard to find maps. 8.5" X 11", 140 ppgs. Retail Price: $11.99

Gold in the Gold Basin and Lost Basin Mining Districts of Mohave County, Arizona - This volume contains rare insights into the geology and gold mineralization of the Gold Basin and Lost Basin Mining Districts of Mohave County, Arizona that will be of benefit to miners and prospectors. Also included is a significant body of information on the gold mines and prospects of this portion of Arizona. This volume is lavishly illustrated with rare photos and mining maps. **8.5" X 11", 188 ppgs. Retail Price: $19.99**

Mines of the Jerome and Bradshaw Mountains of Arizona - This important publication on Arizona Mining has not been available for ninety years. This volume contains rare insights into the geology and ore deposits of the Jerome and Bradshaw Mountains of Arizona that will be of benefit to miners and prospectors who work those areas. Included is a significant body of information on the mines and prospects of the Verde, Black Hills, Cherry Creek, Prescott, Walker, Groom Creek, Hassayampa, Bigbug, Turkey Creek, Agua Fria, Black Canyon, Peck, Tiger, Pine Grove, Bradshaw, Tintop, Humbug and Castle Creek Mining Districts. This volume is lavishly illustrated with rare photos and mining maps. **8.5" X 11", 218 ppgs. Retail Price: $19.99**

The Ajo Mining District of Pima County Arizona - This important publication on Arizona Mining has not been available for nearly seventy years. This volume contains rare insights into the geology and mineralization of the Ajo Mining District in Pima County, Arizona and in particular the famous New Cornelia Mine. **8.5" X 11", 126 ppgs. Retail Price: $11.99**

Mining in the Santa Rita and Patagonia Mountains of Arizona - Originally published in 1915, this important publication on Arizona Mining has not been available for nearly a century. Included are rare insights into hundreds of gold, silver, copper and other mines in this famous Arizona mining area. Details include the locations, geology, history, production and other facts of the mines of this region. **8.5" X 11", 394 ppgs. Retail Price: $24.99**

Mining in the Bisbee Quadrangle of Arizona - Originally published in 1906, this important publication on Arizona Mining has not been available for nearly a century. Included are rare insights into hundreds of gold, silver, copper and other mines in this famous Arizona mining area. Details include the locations, geology, history, production and other facts of the mines of this important mining region. **8.5" X 11", 188 ppgs. Retail Price: $14.99**

Placer Gold Mining in Arizona - Unavailable since 1922, this hard to find publication includes valuable insights into the placer gold mines of the Arizona. Originally released as "Placer Gold of Arizona", despite its small size, this publication includes important details into the history, geology and locations of the well known placer gold mines and prospects in the State of Arizona. **48 ppgs, 8.99**

Gold and Copper Mining near Payson, Arizona - Written in 1915, this hard to find publication includes valuable insights into the gold and copper mining industry of Arizona. Highlighted here are the gold and copper mines near Payson, Arizona. **68 ppgs, 8.99**

Lode Gold Mining in Arizona - Unavailable since 1934, this hard to find publication, originally released as "Arizona Lode Gold Mines and Gold Mining" includes valuable insights into the gold mining industry of Arizona. Included are valuable insights into over 150 hardrock gold mines in over 30 different mining districts in Arizona. **278 ppgs, 21.99**

Mining in the Dragoon Quadrangle of Cochise County, Arizona - Unavailable since 1964, this hard to find publication includes valuable insights into the mines of the Dragoon Quadrangle Mining Region. The publication includes important details into the history, geology and locations of the well known mines and prospects in this famous mining region of Arizona. **224 ppgs., 19.99**

Directory of Operating Mines in Arizona in 1915 - Unavailable since 1916, this hard to find publication includes valuable insights into the mines of Arizona. This small publication includes a complete list of the mines that were operating in the State of Arizona during 1915 and includes details such as general location, owners and some basic facts about each mining operation. **52 ppgs. 8.99**

Arizona Ore Deposits - Unavailable since 1938, this hard to find publication includes valuable insights into some ore deposits of Arizona. Included are valuable insights into the formation and characteristics of valuable ore deposits in the Jerome, Miami, Inspiration, Clifton, Morenci, Ray, Ajo, Eureka, Tombstone and Magma mining districts. Included are details into some of the major gold, silver and copper mines of these important Arizona mining areas. **160 ppgs, 14.99**

Montana Mining Books

A History of Butte Montana: The World's Greatest Mining Camp - First published in 1900 by H.C. Freeman, this important publication sheds a bright light on one of the most important mining areas in the history of The West. Together with his insights, as well as rare photographs of the periods, Harry Freeman describes Butte and its vicinity from its early beginnings, right up to its flush years when copper flowed from its mines like a river. At the time of publication, Butte, Montana was known worldwide as "The Richest Mining Spot On Earth" and produced not only vast amounts of copper, but also silver, gold and other metals from its mines. Freeman illustrates, with great detail, the most important mines in the vicinity of Butte, providing rare details on their owners, their history and most importantly, how the mines operated and how their treasures were extracted. Of particular interest are the dozens of rare photographs that depict mines such as the famous Anaconda, the Silver Bow, the Smoke House, Moose, Paulin, Buffalo, Little Minah, the Mountain Consolidated, West Greyrock, Cora, the Green Mountain, Diamond, Bell, Parnell, the Neversweat, Nipper, Original and many others.   **8.5″ X 11″, 142 ppgs. Retail Price: $12.99**

The Butte Mining District of Montana - This important publication on Montana Mining has not been available for over a century. Included are rare insights into the gold, copper and silver mines of Butte, Montana together with hard to find maps and photographs. Some of the topics include the early history of gold, silver and copper mining in the Butte area, insight into the geology of its mining areas, the local distribution of gold, silver and copper ores, as well their composition and how to identify them. Also included are detailed facts about the mines in the Butte Mining District, including the famous Anaconda Mine, Gagnon, Parrot, Blue Vein, Moscow, Poulin, Stella, Buffalo, Green Mountain, Wake Up Jim, the Diamond-Bell Group, Mountain Consolidated, East Greyrock, West Greyrock, Snowball, Corra, Speculator, Adirondack, Miners Union, the Jessie-Edith May Group, Otisco, Iduna, Colorado, Lizzie, Cambers, Anderson, Hesperus, Preferencia and dozens of others. **8.5″ X 11″, 298 ppgs. Retail Price: $24.99**

Mines of the Helena Mining Region of Montana - This important publication on Montana Mining has not been available for over a century. Included are rare insights into the gold, copper and silver mines of the vicinity of Helena, Montana, including the Marysville Mining District, Elliston Mining District, Rimini Mining District, Helena Mining District, Clancy Mining District, Wickes Mining District, Boulder and Basin Mining Districts and the Elkhorn Mining District. Some of the topics include the early history of gold, silver and copper mining in the Helena area, insight into the geology of its mining areas, the local distribution of gold, silver and copper ores, as well their composition and how to identify them. Also included are detailed facts, history, geology and locations of over one hundred gold, silver and copper mines in the area . **8.5″ X 11″, 162 ppgs, Retail Price: $14.99**

Mines and Geology of the Garnet Range of Montana - This important publication on Montana Mining has not been available for over a century. Included are rare insights into the gold, copper and silver mines of the vicinity of this important mining area of Montana. Some of the topics include the early history of gold, silver and copper mining in the Garnet Mountains, insight into the geology of its mining areas, the local distribution of gold, silver and copper ores, as well their composition and how to identify them. Also included are detailed facts, history, geology and locations of numerous gold, silver and copper mines in the area . **8.5″ X 11″, 100 ppgs, Retail Price: $11.99**

Mines and Geology of the Philipsburg Quadrangle of Montana - This important publication on Montana Mining has not been available for over a century. Included are rare insights into the gold, copper and silver mines of the vicinity of this important mining area of Montana. Some of the topics include the early history of gold, silver and copper mining in the Philipsburg Quadrangle, insight into the geology of its mining areas, the local distribution of gold, silver and copper ores, as well their composition and how to identify them. Also included are detailed facts, history, geology and locations of over one hundred gold, silver and copper mines in the area **8.5″ X 11″, 290 ppgs, Retail Price: $24.99**

Geology of the Marysville Mining District of Montana - Included are rare insights into the mining geology of the Marysville Mining District. Some of the topics include the early history of gold, silver and copper mining in the area, insight into the geology of its mining areas, the local distribution of gold, silver and copper ores, as well their composition and how to identify them. Also included are detailed facts, history, geology and locations of gold, silver and copper mines in the area   **8.5″ X 11″, 198 ppgs, Retail Price: $19.99**

The Geology and Mines of Northern Idaho and North Western Montana- See listing under Idaho.

The History of Gold Dredging in Montana - Unavailable since 1916, this important publication was originally published by the Us Bureau of Mines and has been unavailable for a century. A century and more ago, giant dredging machines dug in Montana's rivers and creeks in search of illusive golden riches. First appearing in California in the 1850's, gold dredges finally reached their peak of development in Siberia and New Zealand before becoming popular again in the United States. This book offers a unique historical perspective on the gold dredges that once operated in Montana. This book on Montana mining history is lavishly illustrated with dozens of rare historic photos gold dredges that once operated in Montana, as well as hard to locate plans on how these dredges were designed. 120 ppgs., 11.99

Nevada Mining Books

The Bull Frog Mining District of Nevada - Unavailable since 1910, this publication was originally compiled by the United States Department of Interior. This volume also includes important insights into the geologic formations, faults and other aspects of economic geology in this Nevada mining district. Of particular interest are the fine details on many mines in the area, including their locations, histories, development and mineralization. Some of the mines featured include the National Bank Mine, Providence, Gibraltor, Tramps, Denver, Original Bullfrog, Gold Bar, Mayflower, Homestake-King and other mines and prospects. **8.5" X 11", 152 ppgs, Retail Price: $14.99**

History of the Comstock Lode - Unavailable since 1876, this publication was originally released by John Wiley & Sons. This volume also includes important insights into the famous Comstock Lode of Nevada that represented the first major silver discovery in the United States. During its spectacular run, the Comstock produced over 192 million ounces of silver and 8.2 million ounces of gold. Not only did the Comstock result in one of the largest mining rushes in history and yield immense fortunes for its owners, but it made important contributions to the development of the State of Nevada, as well as neighboring California. Included here are important details on not only the early development and history of the Comstock, but also rare early insight into its mines, ore and its geology.**8.5" X 11", 244 ppgs, Retail Price: $19.99**

The Pioche Mining District of Nevada - First published in 1932, it has been unavailable for over a century and sheds important light on the mining history of Nevada. Some of the topics include the history of mining in this district, as well as the characteristics of its mineral and ore deposits. Also included are insights into the history, production, characteristics and locations of numerous mines in the area. Some of the mines include the Combined Metals, Pioche, Ely Valley, No. 10, Poorman, Wide Awake, Alps, Prince, Virginia Louise, Half Moon, Abe Lincoln, Fairview, Bristol Silver, National, Vesuvius, Inman, Tempest, Hillside, Jackrabbit, Lucky Star, Fortuna, Mendha, Manhattan, Hamburg, Comet, Lyndon and others. 108 ppgs 10.99

The Yerington Mining District of Nevada - First published in 1932, it has been unavailable for over a century and sheds important light on the mining history of Nevada. Some of the topics include the history of mining in this district, as well as the characteristics of its mineral and ore deposits. Also included are insights into the history, production, characteristics and locations of numerous mines in the area. Some of the mines include the Bluestone, Mason Valley, Malachite, McConnell, Greenwood, Western Nevada, Ludwig, Douglas Hill, Casting Copper, Montana-Yerington, Empire, Jim Beatty, Terry and McFarland, Blue Jay and others. 92 ppgs, 10.99

The Genesis of the Ores of Tonopah Nevada - Unavailable since 1918, this hard to find publication includes valuable insights into the gold mines around Tonopah, Nevada. The publication includes important details into the geology of mines in the Tonopah Mining District of Nevada. 90 ppgs, 10.99

Mining Camps of Elko, Lander and Eureka Counties Nevada - Unavailable since 1910, this hard to find publication includes valuable insights into the mining camps of Elko, Lander and Eureka Counties, Nevada. The publication includes important details into the history of mines and mining in these three Nevada counties. 154 ppgs, 12.99

Ore Deposits of the Bullfrog Quadrangle - Unavailable since 1964 and released as "Geology of Bullfrog Quadrangle and Ore Deposits Related to Bullfrog Hills Caldera, Nye County, Nevada and Inyo County, California". The publication includes important details into the geology of mines in the Bullfrog Quadrangle of Nye County, Nevada and Inyo County, California. 52 ppgs, 9.99

Mining in Eureka County Nevada - Unavailable since 1879, this hard to find publication includes valuable insights into the early mining history off Eureka County, Nevada. The publication includes important details into the early history of the mines of Eureka County, as well as their development, production and how their ores were treated. Also included are details on the 1872 Mining Act, as well as the local rules, regulations and customs of the miners in Eureka County.134 ppgs, 12.99

Colorado Mining Books

Ores of The Leadville Mining District - Unavailable since 1926, this publication was originally compiled by the United States Department of Interior. This volume also includes important insights into the ores and mineralization of the Leadville Mining District in Colorado. Topics include historic ore prospecting methods, local geology, insights into ore veins and stockworks, the local trend and distribution of ore channels, reverse faults, shattered rock above replacement ore bodies, mineral enrichment in oxidized and sulphide zones and more. 8.5" X 11", 66 ppgs, **Retail Price: $8.99**

Mining in Colorado - Unavailable since 1926, this publication was originally compiled by the United States Department of Interior. This volume also includes important insights into the mining history of Colorado from its early beginnings in the 1850's right up to the mid 1920's. Not only is Colorado's gold mining heritage included, but also its silver, copper, lead and zinc mining industry. Each mining area is treated separately, detailing the development of Colorado's mines on a county by county basis. 8.5" X 11", 284 ppgs, **Retail Price: $19.99**

Gold Mining in Gilpin County Colorado - Unavailable since 1876, this publication was originally compiled by the Register Steam Printing House of Central City, Colorado. A rare glimpse at the gold mining history and early mines of Gilpin County, Colorado from their first discovery in the 1850's up to the "flush years" of the mid 1870's. Of particular interest is the history of the discovery of gold in Gilpin County and details about the men who made those first strikes. Special focus is given to the early gold mines and first mining districts of the area, many of which are not detailed in other books on Colorado's gold mining history. 8.5" X 11", 156 ppgs, **Retail Price: $12.99**

Mining in the Gold Brick Mining District of Colorado - Important insights into the history of the Gold Brick Mining District, as well as its local geography and economic geology. Also included are the histories and locations of historic mines in this important Colorado Mining District, including the Cortland, Carter, Raymond, Gold Links, Sacramento, Bassick, Sandy Hook, Chronicle, Grand Prize, Chloride, Granite Mountain, Lucille, Gray Mountain, Hilltop, Maggie Mitchell, Silver Islet, Revenue, Roosevelt, Carbonate King and others. In addition to hardrock mining, are also included are details on gold placer mining in this portion of Colorado. 8.5" X 11", 140 ppgs, **Retail Price: $12.99**

Ore Deposits of the London Fault of Colorado - First published in 1941, it has been unavailable since those days and sheds important light on the mines and mineral deposits of the London Fault in Central Colorado's Alma Mining District. This publication sheds important light on the gold veins and lead-silver deposits of the Alma Mining District. Included are geologic details on the London Mine, American Mine, Havigorst Tunnel, Ophir Mine, Mosher Tunnel, London-Butte Mine, Venture Shaft, Hard-To-Beat Mine, Oliver Twist Tunnel, Sacramento Mine, Mudsill Mine, Sherwood Mine, Wagner, Barcoe Tunnel and other mines in this important mining region. 110 ppgs., 10.99

The Mines of Colorado - First published in 1867, it has been unavailable since those days and sheds important light on Colorado's early mining history. Written shortly after the events took place, this publication sheds important light on the Pike's Peak Gold Rush, the discovery of gold on Ralston Creek and Dry Creek in the 1850's, as well as details on the first wave of miners into Colorado and their trials and tribulations as they crossed the Great Plains. Also included are details on early discoveries of lode gold in the mountainous regions of Colorado, details on the early mines hardrock and placer mines, and much more. It is a veritable treasure trove on Colorado's early mining history and will be of great importance to anyone who is interested in the mining of gold or other minerals in Colorado, as well as those interested in the history of the state. 478 ppgs., 29.99

The La Plata Mining District of Colorado - Originally titled "Geology and Ore Deposits in the Vicinity of the La Plata District of Colorado" and first published in 1949, it has been unavailable since those days and sheds important light on the mines and mineral deposits of the La Plata Mining District of Colorado. 214 ppgs., 19.99

Washington Mining Books

The Republic Mining District of Washington - Unavailable since 1910, this important publication was originally published by the Washington Geologic Survey and has been unavailable for a century. Topics include the geology, rock formations and the formation of ore deposits in this important mining area of Washington State. Also included are hard to find details on the geology, history and locations of dozens of mines in the area. Some of the mines featured include the New Republic Mine, Ben Hur, Morning Glory, the South Republic Mine, Quilp, Surprise, Black Tail, Lone Pine, San Poil, Mountain Lion, Tom Thumb, Elcaliph and many others. **8.5" X 11", 94 ppgs, Retail Price: $10.99**

The Myers Creek and Nighthawk Mining Districts of Washington - Unavailable since 1911, this important publication was originally published by the Washington Geologic Survey and has been unavailable for a century. Topics include the geology, rock formations and the formation of ore deposits in these important mining areas of Washington State. Also included are hard to find details on the geology, history and locations of dozens of mines in the area. Some of the mines featured include the Grant Mine, Monterey, Nip and Tuck, Myers Creek, Number Nine, Neutral, Rainbow, Aztec, Crystal Butte, Apex, Butcher Boy, Molson, Mad River, Olentangy, Delate, Kelsey, Golden Chariot, Okanogan, Ohio, Forty-Ninth Parallel, Nighthawk, Favorite, Little Chopaka, Summit, Number One, California, Peerless, Caaba, Prize Group, Ruby, Mountain Sheep, Golden Zone, Rich Bar, Similkameen, Kimberly, Triune, Hiawatha, Trinity, Hornsilver, Maquae, Bellevue, Bullfrog, Palmer Lake, Ivanhoe, Copper World and many others. **8.5" X 11", 136 ppgs, Retail Price: $12.99**

The Blewett Mining District of Washington - Unavailable since 1911, this important publication was originally published by the Washington Geologic Survey and has been unavailable for a century. Topics include the geology, rock formations and the formation of ore deposits in this important mining area of Washington State. Also included are hard to find details on the geology, history and locations of dozens of mines in the area. Some of the mines featured include the Washington Meteor, Alta Vista, Pole Pick, Blinn, North Star, Golden Eagle, Tip Top, Wilder, Golden Guinea, Lucky Queen, Blue Bell, Prospect, Homestake, Lone Rock, Johnson, and others. **8.5" X 11", 134 ppgs, Retail Price: $12.99**

Silver Mining In Washington - Unavailable since 1955, this important publication was originally published by the Washington Geologic Survey. Featured are the hard to find locations and details pertaining to Washington's silver mines. **8.5" X 11", 180 ppgs, Retail Price: $15.99**

The Mines of Snohomish County Washington - Unavailable since 1942, this important publication was originally published by the Washington Geologic Survey and has been unavailable for seventy years. Featured are details on a large number of gold, silver, copper, lead and other metallic mineral mines. Included are the locations of each historic mine, along with information on the commodity produced. **8.5" X 11", 98 ppgs, Retail Price: $10.99**

The Mines of Chelan County Washington - Unavailable since 1943, this important publication was originally published by the Washington Geologic Survey and has been unavailable for seventy years. Featured are details on a large number of gold, silver, copper, lead and other metallic mineral mines. Included are the locations of each historic mine, along with information on the commodity. **8.5" X 11", 88 ppgs, Retail Price: $9.99**

Metal Mines of Washington - Unavailable since 1921, this important publication was originally published by the Washington Geologic Survey and has been unavailable for nearly ninety years. Widely considered a masterpiece on the Washington Mining Industry, "Metal Mines of Washington" sheds light on the important details of Washington's early mining years. Featured are details on hundreds of gold, silver, copper, lead and other metallic mineral mines. Included are hard to find details on the mineral resources of this state, as well as the locations of historic mines. Lavishly illustrated with maps and historic photos and complete with a glossary to explain any technical terms found in the text, this is one of the most important works on mining in the State of Washington. No prospector or miner should be without it if they are interested in mining in Washington. **8.5" X 11", 396 ppgs, Retail Price: $24.99**

Gem Stones In Washington - Unavailable since 1949, this important publication was originally published by the Washington Geologic Survey and has been unavailable since first published. Included are details on where to find naturally occurring gem stones in the State of Washington, including quartz crystal, amethyst, smoky quartz, milky quartz, agates, bloodstone, carnelian, chert, flint, jasper, onyx, petrified wood, opal, fire opal, hyalite and others. **8.5" X 11", 54 ppgs, Retail Price: $8.99**

The Covada Mining District of Washington - Unavailable since 1913, this important publication was originally published by the Washington Geologic Survey and has been unavailable for a century. Topics include the geology, rock formations and the formation of ore deposits in this important mining area of Washington State. Also included are hard to find details on the geology, history and locations of dozens of mines in the area. Some of the mines featured include the Admiral, Advance, Algonkian, Big Bug, Big Chief, Big Joker, Black Hawk, Black Tail, Black Thorn, Captain, Cherokee Strip, Colorado, Dan Patch, Dead Shot, Etta, Good Ore, Greasy Run, Great Scott, Idora, IXL, Jay Bird, Kentucky Bell, King Solomon, Laurel, Laura S, Little Jay, Meteor, Neglected, Northern Light, Old Nell, Plymouth Rock, Polaris, Quandary, Reserve, Shoo Fly, Silver Plume, Three Pines, Vernie, White Rose and dozens of others. **8.5" X 11", 114 ppgs, Retail Price: $10.99**

The Index Mining District of Washington - Unavailable since 1912, this important publication was originally published by the Washington Geologic Survey and has been unavailable for a century. Topics include the geology, rock formations and the formation of ore deposits in this important mining area of Washington State. Also included are hard to find details on the geology, history and locations of dozens of mines in the area. Some of the mines featured include the Sunset, Non-Pareil, Ethel Consolidated, Kittaning, Merchant, Homestead, Co-operative, Lost Creek, Uncle Sam, Calumet, Florence-Rae, Bitter Creek, Index Peacock, Gunn Peak, Helena, North Star, Buckeye. Copper Bell, Red Cross and others. **8.5" X 11", 114 ppgs, Retail Price: $11.99**

Mining & Mineral Resources of Stevens County Washington - Unavailable since 1920, this important publication was originally published by the Washington Geologic Survey and has been unavailable for a century. Topics include the geology, rock formations and the formation of ore deposits in these important mining areas of Washington State. Also included are hard to find details on the geology, history and locations of hundreds of mines in the area. **8.5" X 11", 372 ppgs, Retail Price: $24.99**

The Mines and Geology of the Loomis Quadrangle Okanogan County, Washington - Unavailable since 1972, this important publication was originally published by the Washington Geologic Survey and has been unavailable for a century. Topics include the geology, rock formations and the formation of ore deposits in this important mining area of Washington State. Also included are hard to find details on the geology, history and locations of dozens of gold, copper, silver and other mines in the area. **8.5" X 11", 150 ppgs, Retail Price: $12.99**

The Conconully Mining District of Okanogan County Washington - Unavailable since 1973, this important publication was originally published by the Washington Geologic Survey and has been unavailable for a century. Topics include the geology, rock formations and the formation of ore deposits in this important mining area of Washington State, which also includes Salmon Creek, Blue Lake and Galena. Also included are hard to find details on the geology, mining history and locations of dozens of mines in the area. Some of the mines include Arlington, Fourth of July, Sonny Boy, First Thought, Last Chance, War Eagle-Peacock, Wheeler, Mohawk, Lone Star, Woo Loo Moo Loo, Keystone, Hughes, Plant-Callahan, Johnny Boy, Leuena, Gubser, John Arthur, Tough Nut, Homestake, Key and many others **8.5" X 11", 68 ppgs, Retail Price: $8.99**

Wyoming Mining Books

Mining in the Laramie Basin of Wyoming - Unavailable since 1909, this publication was originally compiled by the United States Department of Interior. Also included are insights into the mineralization and other characteristics of this important mining region, especially in regards to coal, limestone, gypsum, bentonite clay, cement, sand, clay and copper. **8.5" X 11", 104 ppgs, Retail Price: $11.99**

New Mexico Mining Books

The Mogollon Mining District of New Mexico - Unavailable since 1927, this important publication was originally published by the US Department of Interior and has been unavailable for 80 years. Topics include the geology, rock formations and the formation of ore deposits in this important mining area in New Mexico. Of particular focus is information on the history and production of the ore deposits in this area, their form and structure, vein filling, their paragenesis, origins and ore shoots, as well as oxidation and supergene enrichment. Also included are hard to find details, including the descriptions and locations of numerous gold, silver and other types of mines, including the Eureka, Pacific, South Alpine, Great Western, Enterprise, Buffalo, Mountain View, Floride, Gold Dust, Last Chance, Deadwood, Confidence, Maud S., Deep Down, Little Fanney, Trilby, Johnson, Alberta, Comet, Golden Eagle, Cooney, Queen, the Iron Crown, Eberle, Clifton, Andrew Jackson mine, Mascot and others. **8.5" X 11", 144 ppgs, Retail Price: $12.99**

The Percha Mining District of Kingston New Mexico - Unavailable since 1883, this important publication was originally published by the Kingston Tribune and has been unavailable for over one hundred and thirty five years. Having been written during the earliest years of gold and silver mining in the Percha Mining District, unlike other books on the subject, this work offers the unique perspective of having actually been written while the early mining history of this area was still being made. In fact, the work was written so early in the development of this area that many of the notable mines in the Percha District were less than a few years old and were still being operated by their original discoverers with the same enthusiasm as when they were first located. Included are hard to find details on the very earliest gold and silver mines of this important mining district near Kingston in Sierra County, New Mexico. **8.5" X 11", 68 ppgs, Retail Price: $9.99**

East Coast Mining Books

<u>The Gold Fields of the Southern Appalachians</u> - Unavailable since 1895, this important publication was originally published by the US Department of Interior and has been unavailable for nearly 120 years. Topics include the geology, rock formations and the formation of ore deposits in this important mining area of the American South. Of particular focus is information on the history and statistics of the ore deposits in this area, their form and structure and veins. Also included are details on the placer gold deposits of the region. The gold fields of the Georgian Belt, Carolinian Belt and the South Mountain Mining District of North Carolina are all treated in descriptive detail. Included are hard to find details, including the descriptions and locations of numerous gold mines in Georgia, North Carolina and elsewhere in the American South. Also included are details on the gold belts of the British Maritime Provinces and the Green Mountains. **8.5" X 11", 104 ppgs, Retail Price: $9.99**

Gold Rush Tales Series

<u>Millions in Siskiyou County Gold</u> - In this first volume of the "Gold Rush Tales" series, leading mining historian and editor Kerby Jackson, introduces us to the story of how millions of dollars worth of gold was discovered in Siskiyou County during the California Gold Rush. Lavishly illustrated with photos from the 19th Century, this hard to find information was first published in 1897 and sheds important light onto the gold rush era in Siskiyou County, California and the experiences of the men who dug for the gold and actually found it. **8.5" X 11", 82 ppgs, Retail Price: $9.99**

<u>The California Rand in the Days of '49</u> - In this second volume of the "Gold Rush Tales" series, leading mining historian and editor Kerby Jackson, introduces us to four tales from the California Gold Rush. Lavishly illustrated with photos from the 19th Century, this hard to find information was first published in 1890's and includes the stories of "California's Rand", details about Chinese miners, how one early miner named Baker struck it rich and also the story of Alphonzo Bowers, who invented the first hydraulic gold dredge. **8.5" X 11", 54 ppgs, Retail Price: $9.99**

More Mining Books

<u>Prospecting and Developing A Small Mine</u> - Topics covered include the classification of varying ores, how to take a proper ore sample, the proper reduction of ore samples, alluvial sampling, how to understand geology as it is applied to prospecting and mining, prospecting procedures, methods of ore treatment, the application of drilling and blasting in a small mine and other topics that the small scale miner will find of benefit. **8.5" X 11", 112 ppgs, Retail Price: $11.99**

<u>Timbering For Small Underground Mines</u> - Topics covered include the selection of caps and posts, the treatment of mine timbers, how to install mine timbers, repairing damaged timbers, use of drift supports, headboards, squeeze sets, ore chute construction, mine cribbing, square set timbering methods, the use of steel and concrete sets and other topics that the small underground miner will find of benefit. This volume also includes twenty eight illustrations depicting the proper construction of mine timbering and support systems that greatly enhance the practical usability of the information contained in this small book. **8.5" X 11", 88 ppgs. Retail Price: $10.99**

<u>Timbering and Mining</u> - A classic mining publication on Hard Rock Mining by W.H. Storms. Unavailable since 1909, this rare publication provides an in depth look at American methods of underground mine timbering and mining methods. Topics include the selection and preservation of mine timbers, drifting and drift sets, driving in running ground, structural steel in mine workings, timbering drifts in gravel mines, timbering methods for driving shafts, positioning drill holes in shafts, timbering stations at shafts, drainage, mining large ore bodies by means of open cuts or by the "Glory Hole" system, stoping out ore in flat or low lying veins, use of the "Caving System", stoping in swelling ground, how to stope out large ore bodies, Square Set timbering on the Comstock and its modifications by California miners, the construction of ore chutes, stoping ore bodies by use of the "Block System", how to work dangerous ground, information on the "Delprat System" of stoping without mine timbers, construction and use of headframes and much more. This volume provides a reference into not only practical methods of mining and timbering that may be employed in narrow vein mining by small miners today, but also rare insights into how mines were being worked at the turn of the 19th Century. **8.5" X 11", 288 ppgs. Retail Price: $24.99**

A Study of Ore Deposits For The Practical Miner - Mining historian Kerby Jackson introduces us to a classic mining publication on ore deposits by J.P. Wallace. First published in 1908, it has been unavailable for over a century. Included are important insights into the properties of minerals and their identification, on the occurrence and origin of gold, on gold alloys, insights into gold bearing sulfides such as pyrites and arsenopyrites, on gold bearing vanadium, gold and silver tellurides, lead and mercury tellurides, on silver ores, platinum and iridium, mercury ores, copper ores, lead ores, zinc ores, iron ores, chromium ores, manganese ores, nickel ores, tin ores, tungsten ores and others. Also included are facts regarding rock forming minerals, their composition and occurrences, on igneous, sedimentary, metamorphic and intrusive rocks, as well as how they are geologically disturbed by dikes, flows and faults, as well as the effects of these geologic actions and why they are important to the miner. Written specifically with the common miner and prospector in mind, the book will help to unlock the earth's hidden wealth for you and is written in a simple and concise language that anyone can understand. **8.5″ X 11″, 366 ppgs. Retail Price: $24.99**

Mine Drainage - Unavailable since 1896, this rare publication provides an in depth look at American methods of underground mine drainage and mining pump systems. This volume provides a reference into not only practical methods of mining drainage that may be employed in narrow vein mining by small miners today, but also rare insights into how mines were being worked at the turn of the 19th Century. **8.5″ X 11″, 218 ppgs. Retail Price: $24.99**

Fire Assaying Gold, Silver and Lead Ores - Unavailable since 1907, this important publication was originally published by the Mining and Scientific Press and was designed to introduce miners and prospectors of gold, silver and lead to the art of fire assaying. Topics include the fire assaying of ores and products containing gold, silver and lead; the sampling and preparation of ore for an assay; care of the assay office, assay furnaces; crucibles and scorifiers; assay balances; metallic ores; scorification assays; cupelling; parting' crucible assays, the roasting of ores and more. This classic provides a time honored method of assaying put forward in a clear, concise and easy to understand language that will make it a benefit to even beginners. **8.5″ X 11″, 96 ppgs. Retail Price: $11.99**

Methods of Mine Timbering - Originally published in 1896, this important publication on mining engineering has not been available for nearly a century. Included are rare insights into historical methods of timbering structural support that were used in underground metal mines during the California that still have a practical application for the small scale hardrock miner of today. **8.5″ X 11″, 94 ppgs. Retail Price: $10.99**

The Enrichment of Copper Sulfide Ores - First published in 1913, it has been unavailable for over a century. Topics include the definition and types of ore enrichment, the oxidation of copper ores, the precipitation of metallic sulfides. Also included are the results of dozens of lab experiments pertaining to the enrichment of sulfide ores that will be of interest to the practical hard rock mine operator in his efforts to release the metallic bounty from his mine's ore. **8.5″ X 11″, 92 ppgs. Retail Price: $9.99**

A Study of Magmatic Sulfide Ores - Unavailable since 1914, this rare publication provides an in depth look at magmatic sulfide ores. Some of the topics included are the definition and classification of magmatic ores, descriptions of some magmatic sulfide ore deposits known at the time of publication including copper and nickel bearing pyrrohitic ore bodies, chalcopyrite-bornite deposits, pyritic deposits, magnetite-ileminite deposits, chromite deposits and magmatic iron ore deposits. Also included are details on how to recognize these types of ore deposits while prospecting for valuable hardrock minerals. **8.5″ X 11″, 138 ppgs. Retail Price: $11.99**

The Cyanide Process of Gold Recovery - Unavailable since 1894 and released under the name "The Cyanide Process: Its Practical Application and Economical Results", this rare publication provides an in depth look at the early use of cyanide leaching for gold recovery from hardrock mine ores. This volume provides a reference into the early development and use of cyanide leaching to recover gold. **8.5″ X 11″, 162 ppgs. Retail Price: $14.99**

California Gold Milling Practices - Unavailable since 1895 and released under the name "California Gold Practices", this rare publication provides an in depth look at early methods of milling used to reduce gold ores in California during the late 19th century. This volume provides a reference into the early development and use of milling equipment during the earliest years of the California Gold Rush up to the age of the Industrial Revolution. Much of the information still applies today and will be of use to small scale miners engaging in hardrock mining. **8.5″ X 11″, 104 ppgs. Retail Price: $10.99**

Leaching Gold and Silver Ores With The Plattner and Kiss Processes - Mining historian Kerby Jackson introduces us to a classic mining publication on the evaluation and examination of mines and prospects by C.H. Aaron. First published in 1881, it has been unavailable for over a century and sheds important light on the leaching of gold and silver ores with the Plattner and Kiss processes. **8.5″ X 11″, 204 ppgs. Retail Price: $15.99**

The Metallurgy of Lead and the Desilverization of Base Bullion - First published in 1896, it has been unavailable for over a century and sheds important light on the the recovery of silver from lead based ores. Some of the topics include the properties of lead and some of its compounds, lead ores such as galenite, anglesite, cerussite and others, the distribution of lead ores throughout the United States and the sampling and assaying of lead ores. Also covered is the metallurgical treatment of lead ores, as well as the desilverization of lead by the Pattinson Process and the Parkes Process. Hofman's text has long been considered one of the most important early works on the recovery of silver from lead based ores. 8.5" X 11", 452 ppgs. Retail Price: $29.99

Ore Sampling For Small Scale Miners - First published in 1916, it has been unavailable for over a century and sheds important light on historic methods of ore sampling in hardrock mines. Topics include how to take correct ore samples and the conditions that affect sampling, such as their subdivision and uniformity. Particular detail is given to methods of hand sampling ore bodies by grab sample, pipe sample and coning, as well as sampling by mechanical methods. Also given are insights into the screening, drying and grinding processes to achieve the most consistent sample results and much more. 8.5" X 11", 124 ppgs. Retail Price: $12.99

The Extraction of Silver, Copper and Tin from Ores - First published in 1896, it has been unavailable for over a century and sheds important light on how historic miners recovered silver, copper and tin from their mining operations. The book is split into three sections, including a discussion on the Lixiviation of Silver Ores, the mining and treatment of copper ores as practiced at Tharsis, Spain and the smelting of tin as it was practiced by metallurgists at Pulo Brani, Singapore. Also included is an overview and analysis of these historic metal recovery methods that will be of benefit to those interested in the extraction of silver, copper and tin from small mines. 8.5" X 11", 118 ppgs. Retail Price: $14.99

The Roasting of Gold and Silver Ores - First published in 1880, it has been unavailable for over a century and sheds important light on how historic miners recovered gold and silver rom their mining operations. Topics include details on the most important silver and free milling gold ores, methods of desulphurization of ores, methods of deoxidation, the chlorination of ores, methods and details on roasting gold and silver ores, notes on furnaces and more. Also included are details on numerous methods of gold and silver recovery, including the Ottokar Hofman's Process, the Patera Process, Kiss Process, Augustin Process, Ziervogel Process and others. 8.5" X 11", 178 ppgs. Retail Price: $19.99

The Examination of Mines and Prospects - First published in 1912, it has been unavailable for over a century and sheds important light on how to examine and evaluate hardrock mines, prospects and lode mining claims. Sections include Mining Examinations, Structural Geology, Structural Features of Ore Deposits, Primary Ores and their Distribution, Types of Primary Ore Deposits, Primary Ore Shoots, The Primary Alteration of Wall Rocks, Alterations by Surface Agencies, Residual Ores and their Distribution, Secondary Ores and Ore Shoots and Vein Outcrops. This hard to find information is a must for those who are interested in owning a mine or who already own a lode mining claim and wish to succeed at quartz mining. 8.5" X 11", 250 ppgs. Retail Price: $19.99

Garnets: Their Mining, Milling and Utilization - First published in 1925, it has been unavailable since those days and sheds important light on the mining, milling and utilization of garnets. Included are details on the characteristics of garnets, where they are found and how they were mined. 78 ppgs, 10.99

Gemstones and Precious Stones of North America - Leading mining historian Kerby Jackson introduces us to a classic mining publication on the gems and precious stones of the United States, Canada and mexico. First published in 1890, it has been unavailable since those days and sheds important light on the gems and precious stones that may be found in North America. Included are chapters on diamonds, corundum, sapphire, ruby, topaz, emerald, disapore, spinel, turquoise, tourmaline, garnets, beyrl, peridot, zircon, quartz crystals, feldspars, pearls and many others. Included are details on where these gems and precious stones may be found throughout North America, as well as their characteristics. 360 ppgs, 24.99

Mining Camps and Mining Districts - First released in 1885 by Charles Howard Shinn under the title "Mining Camps: A Study in American Frontier Government", this publication offers a unique look at how early gold miners established their own forms of representative government during the California Gold Rush. Drawing on the the early mining codes of medieval German miners in the Harz Mountains, on the mining customs of the Cornish tin miners and early Spanish mining laws introduced into California, the miners established the first governments in the American West. 340 ppgs, 24.99

BLM Field Handbook for Mineral Examiners - Leading mining historian Kerby Jackson introduces us to a classic mining publication on mine evaluation. First published in 1962, this work sheds important light on the techniques of BLM Mineral Examiners to perform validity on mining claims. 132 ppgs, 10.99

Six Months In The Gold Mines During The California Gold Rush - Unavailable since 1850, this important work is a first hand account of one "49'ers" personal experience during the great California Gold Rush, shedding important light on one of the most exciting periods in the history of not only California, but also the world. Compiled from journals written between 1847 and 1849 by E. Gould Buffum, a native of New York, "Six Months In The Gold Mines During The California Gold Rush" offers a rare look into the day to day lives of the people who came to California to work in her gold mines when the state was still a great frontier. 8.5" X 11", 290 ppgs. Retail Price: $19.99

The Discovery of Gold in Australia - First published in 1852, it has been unavailable since those days and sheds important light on Australia's gold mining history. Included are rare communications between British agents and the British Crown when gold was first discovered in Australia in 1851. This rare text contains hard to find details on Australia's first mining camps and Britain's early attempts to provide for the orderly regulation of gold mines in that part of the world. Also of interest are hard to find extracts of articles that appeared in the early colonial newspapers that did their best to report on Australia's gold rush as it took place.
102 ppgs, 10.99